SEA-FLOOR SEDIMENT

And the Age of the Earth

Larry Vardiman, Ph.D.

First Printing 1996

Sea-Floor Sediment and the Age of the Earth

Institute for Creation Research
P.O. Box 2667
El Cajon, California 92021

Library of Congress Catalog Card Number: 95-081876
ISBN 0-932766-38-2

Cataloging in Publication Data

Printed in the United States of America

ACKNOWLEDGEMENTS

This monograph is dedicated to my daughter, Laura, the youngest of my children. She graduated from high school with honors the year this monograph was completed. I am proud of her accomplishments and anticipate that the Lord will use her for great things in His kingdom.

Thanks to the reviewers who helped make this a better document, especially Gerald Aardsma, John Baumgardner, Richard Bliss, Robert Brown, David Bowdle, Jim Cook, Henry Morris, John Morris, Michael Oard, Andrew Snelling, and Kurt Wise. Any errors still remaining are, of course, mine.

Data for the Deep Sea Drilling Project (DSDP) and the Ocean Drilling Project (ODP) were provided on CD-Rom and diskette by the National Geophysical Data Center (NGDC) Data & Information Service. References to specific reports and data are made in the monograph to the Initial Reports of the DSDP. Analyses were partially conducted on computer equipment provided by Steve Low and his associates with the Hewlett-Packard Company.

I thank Dr. Henry Morris and ICR for providing the opportunity and facilities to conduct the research supporting this monograph. It was a very real joy to be able to work on this project. The opportunity to "... think God's thoughts after Him..." is not available to everyone.

TABLE OF CONTENTS

LIST OF FIGURES

LIST OF TABLES

CHAPTER 1

INTRODUCTION

It is common knowledge that near the mouth of a muddy river flowing into the ocean, such as shown in Fig. 1.1, sediments transported by the river slowly settle out of the water and form deposits on the sea floor. In some locations, such as the delta regions near the mouth of the Mississippi River or the Nile River, the build-up of sediments has resulted in the addition of large regions of new land. However, it is less well known that the growth, death, and deposition of microorganisms in the ocean have contributed to the formation of sea-floor sediments, particularly in mid-ocean regions. These microorganisms make up the bulk of what is called plankton.

Fig. 1.1 *Mud and sand eroded from the surrounding land moving toward the sea in South America.*

Sediments, derived from rocks (lithogenous) and various life forms (biogenous), accumulate on the ocean floor and form a record of earth history (See Fig. 1.2). If the characteristics of the sediments can be related to events and processes which supplied the sediment, they can be a valuable tool to study earth history. Appendix A discusses the composition, sources, and distribution of sea-floor sediments around the globe.

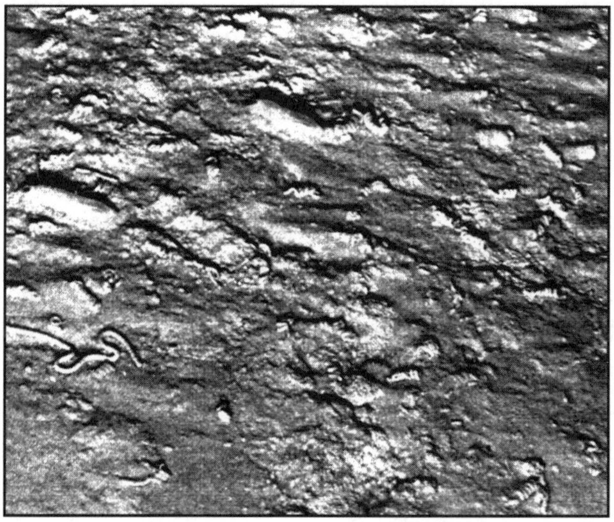

Fig. 1.2 *Sediment on the sea floor at 3,250 meters beneath the circumpolar current in the South Pacific. Note two sea cucumbers in upper left and trackway in middle left.*

Research on sea-floor sediments has been actively pursued for over forty years. Sediment cores have been extracted from the sea floor at locations throughout the earth and analyzed for types of lithogenous material, types of biogenous forms, sedimentation rate, thickness, date of accumulation, and many other interesting features. Appendix B describes the equipment and procedures used to extract sediment cores from the sea floor. Drilling locations for the Deep Sea Drilling Project and the Ocean Drilling Project are also tabulated.

One of the most interesting fields of research has been the study of paleoclimates using the measurement of oxygen isotopes in the shells from types of microorganisms called forams. This specialized field has developed an explanation for climate fluctuations from warm periods in the Cretaceous, when dinosaurs are thought to have roamed the Earth, to cold periods, such as the recent "Ice Age." Strong attempts have been made to explain the cyclical layering of sediments as caused by periodic occurrences of "Ice Ages" caused, in turn, by orbitally-induced fluctuations in solar heating of the earth. Appendix C describes how oxygen isotopes are analyzed in sea-floor sediments to estimate ocean temperatures.

The time frame offered by the conventional explanations of climate suggest that the ocean sediments accumulated over hundreds of millions of years, and recent "Ice Ages" occurred over periods of time on the order of a hundred millennia. These ages are not compatible with a literal interpretation of the Biblical account of creation and earth history. The main sources of disagreement between the conventional model of earth history and a model consistent with the Bible for sediment accumulation, to be presented in this monograph, are the assumptions about the magnitude of the driving mechanism and the process rates. The conventional model assumes sediment accumulated slowly over long periods of time by low-energy processes. The creation model, to be developed in this effort, assumes sediment accumulated rapidly over a relatively short period of time by catastrophic processes during and following the global Flood described in Genesis. Appendix D provides a brief overview of the conventional geological time scale for those readers who are unfamiliar with the nomenclature and time frame. Appendix E surveys the conventional explanations for the estimated temperature change from the Cretaceous period to the present. Quaternary temperature fluctuations are discussed in some depth, particularly in relation to the astronomical theory.

Chapter 2 briefly reviews a literal Biblical chronology and develops a justification for reinterpreting the sea-floor sediment data in terms of a young-earth, catastrophic model. Chapter 3 develops an analytic, young-earth accumulation model of sea-floor sediments. Chapter 4 briefly discusses the implications of a young-earth age model, particularly on the study of climate. Chapter 5 summarizes the results of this effort, presents conclusions, and offers recommendations concerning the new model.

The age of the earth is not derived from this study, but rather, assumed. In fact, the global, catastrophic Flood, described in Genesis, is assumed to have occurred about 4,500 years ago and to have caused world-wide devastation and massive sources of lithogenous and biogenous sediments in the ocean. The accumulation of sea-floor sediments in such a short time results in an entirely new view of earth history. It did not take 65 millions years for the Cretaceous ocean to cool some 15°C to the present — it took only 4,500 years. The last "Ice Age" did not last 100,000 years and was not preceded by 15-30 other "Ice Ages" of similar duration, but, rather, occurred recently following the Flood described in Genesis.

CHAPTER 2

BIBLICAL TIME CONSTRAINTS

The Bible does not speak directly about sea-floor sediments or forams. Nowhere do the Scriptures describe the vast layers of sediment which cover the ocean floor, nor do they discuss the processes by which they were formed. Scripture contains only brief, general references that discuss the creation of the sea and God's control over its devastating power. For example:

> And God said, Let the waters under the heaven be gathered together unto one place, and let the dry land appear: and it was so. And God called the dry land Earth; and the gathering together of the waters called He seas: and God saw that it was good (Genesis 1:9,10).

> Hast thou entered into the springs of the sea? or hast thou walked in the search of the depth (Job 38:16)?

> Fear ye not me? saith the LORD: will ye not tremble at my presence, which have placed the sand for the bound of the sea by a perpetual decree, that it cannot pass it: and though the waves thereof toss themselves, yet can they not prevail; though they roar, yet can they not pass over it (Jeremiah 5:22)?

Yet, it is evident that if a global Flood occurred as described in the Scriptures, catastrophic processes would have occurred and massive quantities of sediments would have been produced and distributed over the continents and the ocean floor. Some sediments may have originated on the third day of the creation week when the continents were separated from the oceans, as described in Genesis 1:9,10. However, it is likely that most of the sediments were produced during the Flood. A description of the Flood events are found in Genesis 7:

> In the six hundredth year of Noah's life, in the second month, the seventeenth day of the month, the same day were all the fountains of the great deep broken up, and the windows of heaven were opened. And the rain was upon the earth forty days and forty nights (Genesis 7:11,12).

> And the flood was forty days upon the earth; and the waters increased, and bare up the ark, and it was lift up above the earth. And the waters prevailed, and were increased greatly upon the earth; and the ark went upon the face of the waters. And the waters prevailed exceedingly upon the earth; and all the high hills, that were under the whole heaven, were covered. Fifteen cubits upward did the waters prevail; and the mountains were covered. And all flesh died that moved upon the earth, both of fowl, and of cattle, and of beast, and every creeping thing that creepeth upon the earth, and every man: All in whose nostrils was the breath of life, of all that was in the dry land, died. And every living

substance was destroyed which was upon the face of the ground, both man, and cattle, and the creeping things, and the fowl of the heaven; and they were destroyed from the earth: and Noah only remained alive, and they that were with him in the ark. And the waters prevailed upon the earth an hundred and fifty days (Genesis 7:17-24).

The Flood is described in Genesis 7 primarily in relation to the destruction of life upon the earth. God's concern centers around man and other life forms. However, if "... every living substance was destroyed which was upon the face of the ground ..." and "... all the high hills, that were under the whole heaven, were covered ...," it is logical to assume that major devastation to the crust of the earth occurred as well. The Scriptures do not address these effects, but if one accepts the Biblical account that a global Flood occurred, then the geologic evidence over the earth bears silent testimony to the destructive power of the Flood.

The timing of sea-floor sediment accumulation according to a Biblical framework hinges on the historicity of the great Flood. According to Scripture, the great Deluge in the days of Noah was a worldwide catastrophe in which "... the world that then was, being overflowed with water, perished" (II Peter 3:6). That the Bible teaches a global flood rather than a local or regional flood is evident for many reasons, among which are the following, according to Morris (1984):

1) The flood waters covered all the high mountains (Gen. 7:19-20) and continued to cover them completely for about nine months (Gen. 8:5). These facts can answer hydraulically to a worldwide flood and to nothing else.

2) Expressions of universality in the account (Gen. 6-9) are not scattered and incidental (as is the case elsewhere in Scripture when apparently universal terms are used in a limited sense), but are repeated and emphasized again and again, constituting the very essence of the narrative. There are at least thirty times in which this universality ("all flesh," "every living thing," "all the high hills under the whole heaven," etc.) is mentioned in these chapters.

3) The worldwide character of the Flood is also assumed in later parts of Scripture. See especially the testimony of the psalmist (Psalm 104:6), of Peter (I Peter 3:20; II Peter 2:5; 3:5-6), and of the Lord Jesus Christ (Matthew 24:37-39).

4) The primary purpose of the Flood was to destroy all mankind. This is seen not only in the numerous statements in Genesis to that effect but also in those of Peter (II Peter 2:5) and of Christ (Luke 17:26-27). This could never have been accomplished by anything less than a global catastrophe.

5) The tremendous size of the ark (which, according to the most conservative calculations, had a volumetric capacity equivalent to that of over five hundred standard railroad stock cars) is an eloquent witness that far more than a regional fauna was to be preserved therein. Its purpose was "to keep seed alive upon the face of all the earth" (Genesis 7:3), quite a pointless provision if the Deluge was local.

6) There would obviously have been no need for an ark at all if the Flood were anything other than universal. Noah and his family could far more easily have migrated to some distant land during the 120 years it took to build the ark. Similarly, the birds and animals of the region could much more simply have been preserved by a process of migration. The Flood narrative is thus made entirely ridiculous by the local-flood hypothesis.

7) God's thrice-repeated promise (Gen. 8:21; 9:11,15) never again to "smite every thing living" by a flood clearly applies only to a universal catastrophe. If the promise referred only to a local flood, it has been repeatedly broken every time there has been a destructive flood anywhere in the world. The local flood notion therefore not only charges Scripture with error, but maintains that God does not keep His promises!

The seven reasons listed above are only a few out of many. In his commentary on Genesis, Morris (1976) has given one hundred reasons why the Flood must be regarded as global.

If one accepts the historicity of the great Flood and its devastation of the earth in the recent past, then how does one understand the sea-floor sediment data to fit within a Biblical framework? The specifics of this reinterpretation will be dealt with in Chapter 3 of this monograph. Here we will provide an overview and introduce the Biblical time constraints.

It is evident from geology that if the Flood was global and recent, it would not have been a tranquil event. The inference is that the Flood waters moved across the surface of the earth and eroded miles of the surface sediments and rocks into fine sands and muds. These sediments were deposited primarily on what are continents today. Only minor quantities of sediment were deposited on the ocean floor compared to the continents. Most of the lower sediments in the geologic column beneath the oceans are completely missing. Volcanoes, eruptions of magma, earthquakes, continental movements, tidal waves, continual rain, mountain building, and massive erosion of canyons and plains contributed to the devastation. Forests were destroyed, vegetation and animals were buried, and massive fossil graveyards were created all over the earth as the waters subsided and the suspended sands and muds began to settle out of the seas.

good writing

As the continents rose and the sea floors sank at the end of the Flood to form a new isostatic adjustment of the earth's crust, some of the unconsolidated sediments were eroded off the continents into the oceans and settled on the floor close to the continental boundaries. The continental sediments solidified into rock (lithification) forming the sedimentary rocks observed all over the earth today. On the ocean floor the deeper sediments lithified under the influence of higher pressures and temperatures which removed or chemically altered the water present in the sediments. The ocean floor is much thinner (5-10 kilometers thick) and therefore closer to the heat of the mantle than the continents (35-60 kilometers thick). Water was drained from the sediments on the continents by gravity, aiding the lithification. The water remaining in the oceans may have been chemically altered as well.

The upper layers of sediment on the ocean floor were formed after the Flood as warm, nutrient-rich oceans "bloomed" with forams, radiolaria, coccoliths, diatoms, and many other forms of microscopic and macroscopic life. The ocean floor was littered with shells and debris of life forms living, growing, and dying in abundance. This debris was mixed with residual sediments coming from the continents. The continents probably contributed large quantities of lithogenous sediments immediately after the Flood because of barren surfaces, unlithified sediments, and greater precipitation than today. The amount of sediments entering and forming in the ocean decreased markedly with time after the Flood. It is possible that the warmth of the ocean and possible heating of the floor from below caused rapid oxidation which produced the red and brown muds.

The conventional old-earth model assigns an age of about 65 million years BP to the end of the Cretaceous period. A literal interpretation of Scripture would suggest that the origin of planet earth occurred quite recently — much less than 65 million years ago. The recent-creation model, which I will use, assumes God created the world in a supernatural creative event some 6,000 years ago, and judged his creation through a worldwide catastrophic Flood some 4,500 years ago. The assumption that the Flood occurred about 4,500 years ago is derived from the Ussher chronology using the Textus Receptus. Some would choose a longer chronology based on the Septuagint and relaxation of additional time constraints (Aardsma, 1993). However, the author prefers this time frame, at least to start the study. Between God's supernatural interventions in the affairs of the world, He normally allows the physical processes to operate according to the laws of science. We wish to determine whether the sea-floor sediment data can be reasonably explained within this conceptual framework.

CHAPTER 3

A YOUNG-EARTH MODEL OF SEA-FLOOR SEDIMENT ACCUMULATION

Introduction

Although a large number of scientists from the conventional, old-earth perspective have investigated sediments on the sea floor and have attempted to develop models for the accumulation rates and associated climates, very few creationists have addressed the issue. One of the few to publish in this field is Roth (1985). He reviewed the field and concluded that:

> ...the thickness of the layers of microscopic shells found on the floor of the ocean is much less than proposed earlier. Present rates of production appear relatively slow, while the biological potential for the rapid production of these shells is tremendous. Limiting factors for rapid production such as paucity of carbonate sources and nutrients could be obviated by a worldwide catastrophe such as the flood described in Scripture. Information is meager and some of it of poor quality. Because of these factors the biblical model of origins does not seem to be invalidated on the basis of our present knowledge about the microscopic shells found on the floor of the ocean.

I will take the same general approach as Roth (1985) in developing a young-earth model of sea-floor accumulation, but will attempt to quantify the results by deriving an analytical model for the variation of $\delta^{18}O$ as a function of time. This will permit inferences to be made about temperature change and accumulation rates.

Depth of Sediments

The occurrence of a global Flood, as described in the Bible, would have produced layers of sediment on both the continents and the sea floor. Many of these sediments would have been laid down rapidly during and immediately following the Flood. After the Flood, as the frequency and intensity of the tectonic events subsided, the rate of lithogenous sediment formation would have decreased in proportion to the decrease in tectonic activity and in proportion to the reestablishment of vegetative cover. Because the oceans would have been well-mixed by the Flood and probably warmed significantly by the energy released from frictional forces and heat from magma, brines, etc., brought up from deep within the earth associated with "... all the fountains of the great deep ... (Gen. 7:11), as well as volcanism, it is likely that biogenic sediment would have increased after the Flood for some time until the nutrients were depleted. As the nutrients were depleted and the ocean cooled and stratified, the biogenic sediments would have decreased with time.

The functional change in sediment formation after the Flood is unknown. However, it is reasonable to assume an exponential decrease in tectonic activity and, consequently, an

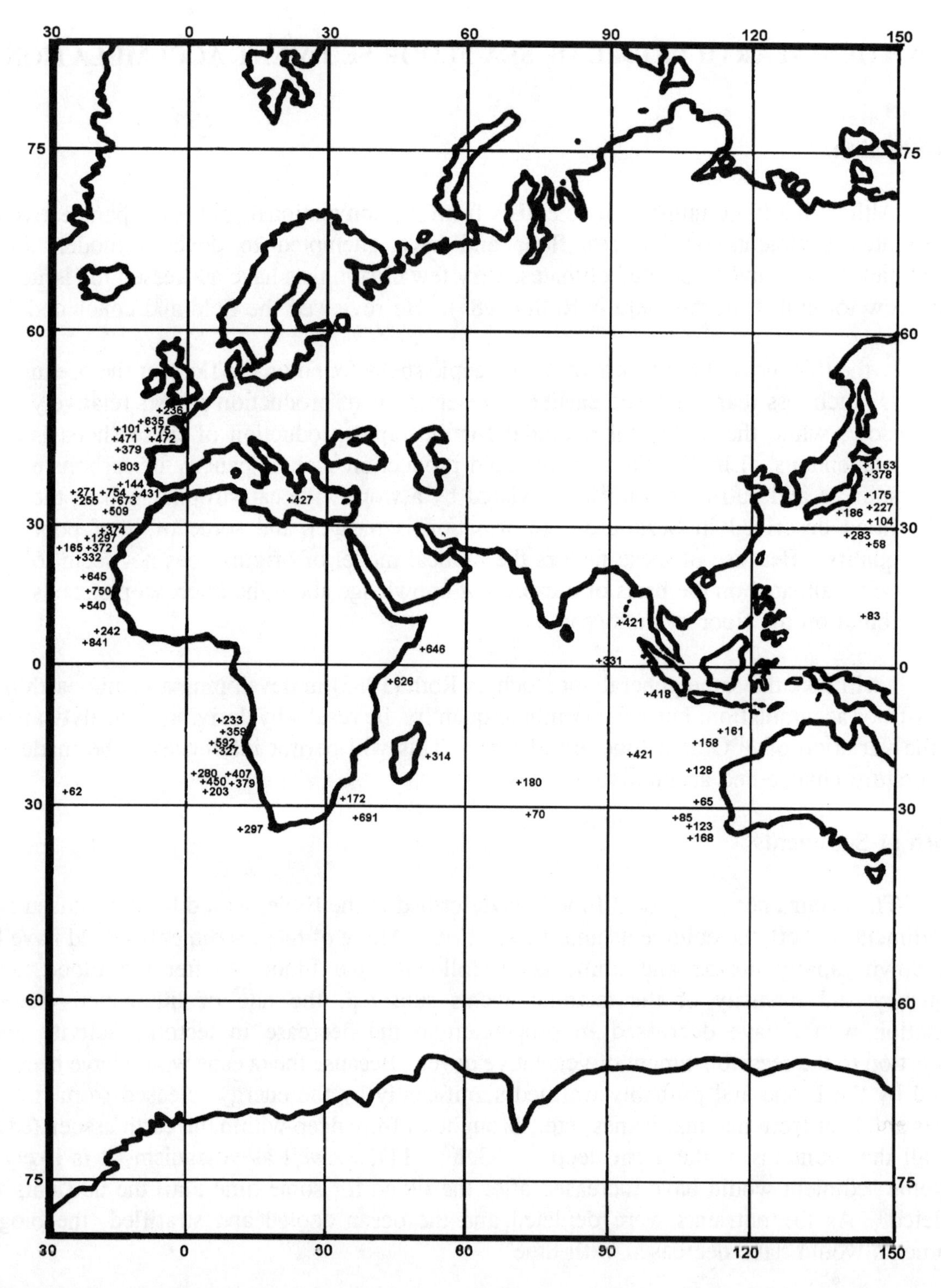

Fig. 3.1 *Thickness of sediments (meters above Cretaceous/Tertiary boundary) for those DSDP sites which contain Cretaceous sediments.*

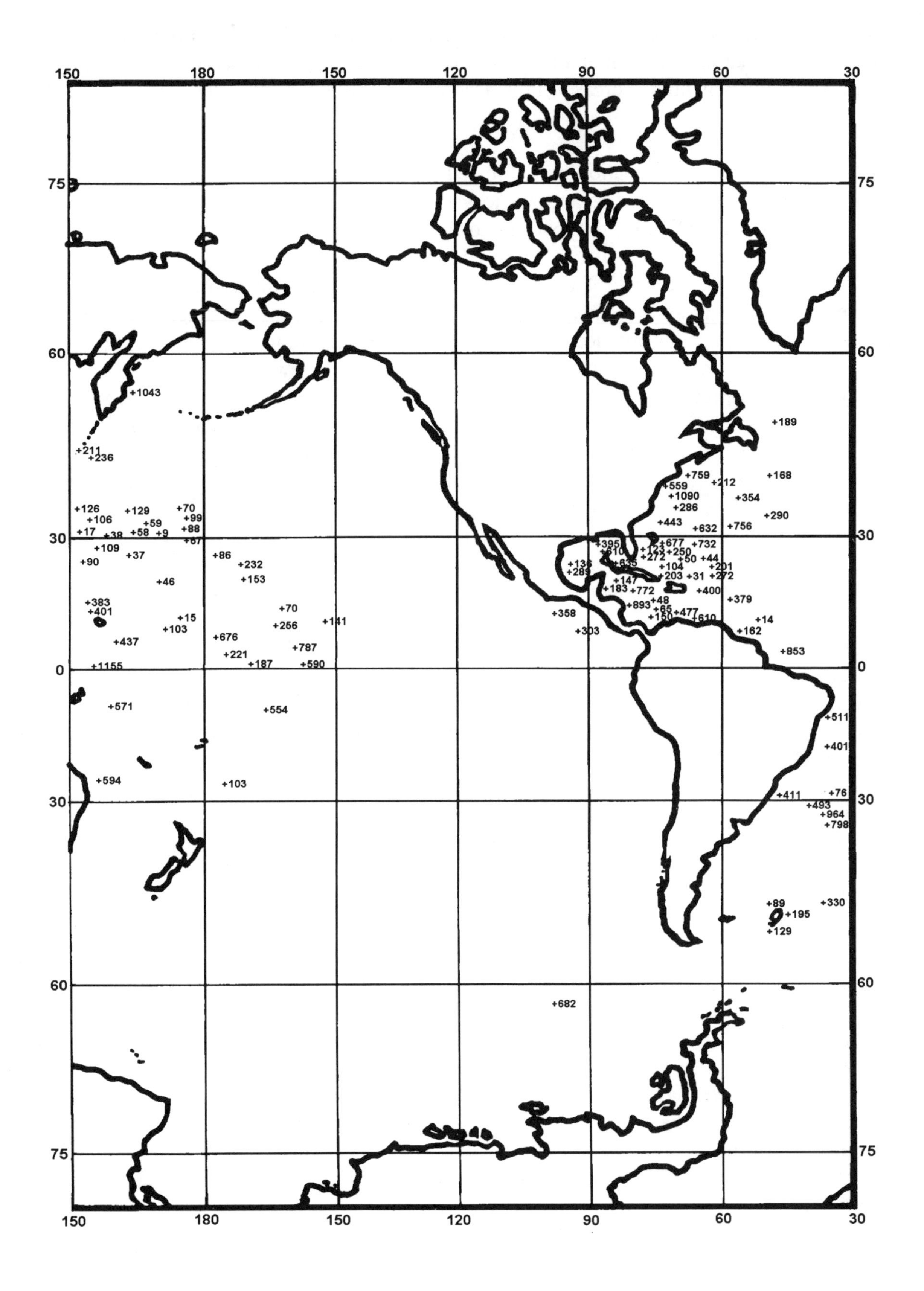

exponential decrease in sediment formation. A refinement to this assumption may need to be made later by treating the accumulation of lithogenous and biogenous sediments separately. For now, a simple exponential decrease in the accumulation rate of sediments, irrespective of type, will be assumed.

The current accumulation rate for sediment formation in the deep ocean has been measured extensively. The rate appears to vary between about 1 cm/1000 years to about 10 cm/1000 years, depending on the investigator and location on the earth. The rate is so small that direct measurements are difficult. In addition, corrections must be made to account for dissolution and other effects. Traps are positioned at various levels in the ocean to collect samples of sediments as they drift downward from biogenous and lithogenous sources. For calibration purposes a uniform accumulation rate is assumed and the observations are compared with the upper layers of sediment formed over the past few hundred years. Since the conventional interpretation of sea-floor sediment accumulation requires millions of years for the formation of the observed layers, it is likely that the average accumulation rates quoted are biased to small values. Nevertheless, the model developed here will assume today's average accumulation rate of deep sea-floor sediment is 2 cm/1000 years or 2×10^{-5} meters/year.

The thickness of sea-floor sediment accumulated since the Flood is unknown. It is unclear how much of the sediment was formed during the energetic events of the Flood and how much formed later as the effects of the Flood subsided. There is no uniformity of opinion among creationists as to the location of the boundary between pre-Flood and Flood rocks on the continents, let alone between Flood and post-Flood strata on the ocean floor. For example, some creationist scientists believe the boundary between pre-Flood and Flood rocks in the Grand Canyon occurs between the Vishnu Schist/Zoroaster Granite and the Tapeats Sandstone at the Great Unconformity about 4,000 feet below the south rim. Others would include the tilted layers of Dox Sandstone, Shinumu Quartzite, Hakatai Shale, and Bass Limestone in the Flood sediments. Some would even include the metamorphosed Vishnu Schist and Zoroaster Granite as Flood layers. Morris (1996) indicates that the entire continental Tertiary Period was probably produced by the events of the Flood. If creationists can't agree on the location of the boundaries between major events on the continents where there are numerous exposures to study, how much less likely is agreement on boundaries in sediments miles under the ocean.

For the purpose of this first study, a partition between the Flood and post-Flood events will be assumed at the Cretaceous/Tertiary boundary, although no sudden global change would have been anticipated as the Flood subsided. This is one of the most recognizable levels in the geologic column. It is the boundary between two of the major eras — the Mesozoic and the Cenozoic. It has been identified by creationists and non-creationists alike as the location of major changes in geologic history. In fact, some evolutionists are now suggesting worldwide catastrophic events at the Cretaceous/Tertiary boundary — namely, the impact of asteroids on the earth, a worldwide dust cloud, global winter, and the destruction of the dinosaurs and many major life forms. Many of these scenarios fit well with the devastation suggested by creationists in the global Flood of Genesis.

In addition to this easily-recognizable boundary and the catastrophism associated with it, the temperatures inferred by the $\delta^{18}O$ record show a decline to the present from a maximum during the Cretaceous Period. If the oceans were heated by events of the Flood, the Cretaceous Period would logically be included in the Flood. Several warm events occurred following the Cretaceous but these were of smaller magnitude, lending support to the idea of the Tertiary coming after the year of the Flood.

The Deep Sea Drilling Project extracted cores from 624 sites from the ocean floors of the globe. Cores from most of these sites showed only recent sediments from the Tertiary and Quaternary periods. Of the 624 total sites only 186 contained sediments from the Cretaceous period or earlier. This means that the ocean floor is relatively young compared to the continents.

For each of the 186 sites containing Cretaceous sediments the thickness of overlying sediment was tabulated. Figure 3.1 shows the thickness of sediment overlying the Cretaceous/Tertiary boundary on a global map. Note that no sites contained Cretaceous sediments along the mid-Atlantic ridge or along the eastern Pacific rise. Apparently, older sediments in the mid-Atlantic and eastern Pacific were never deposited or were subducted during "recent" tectonic processes.

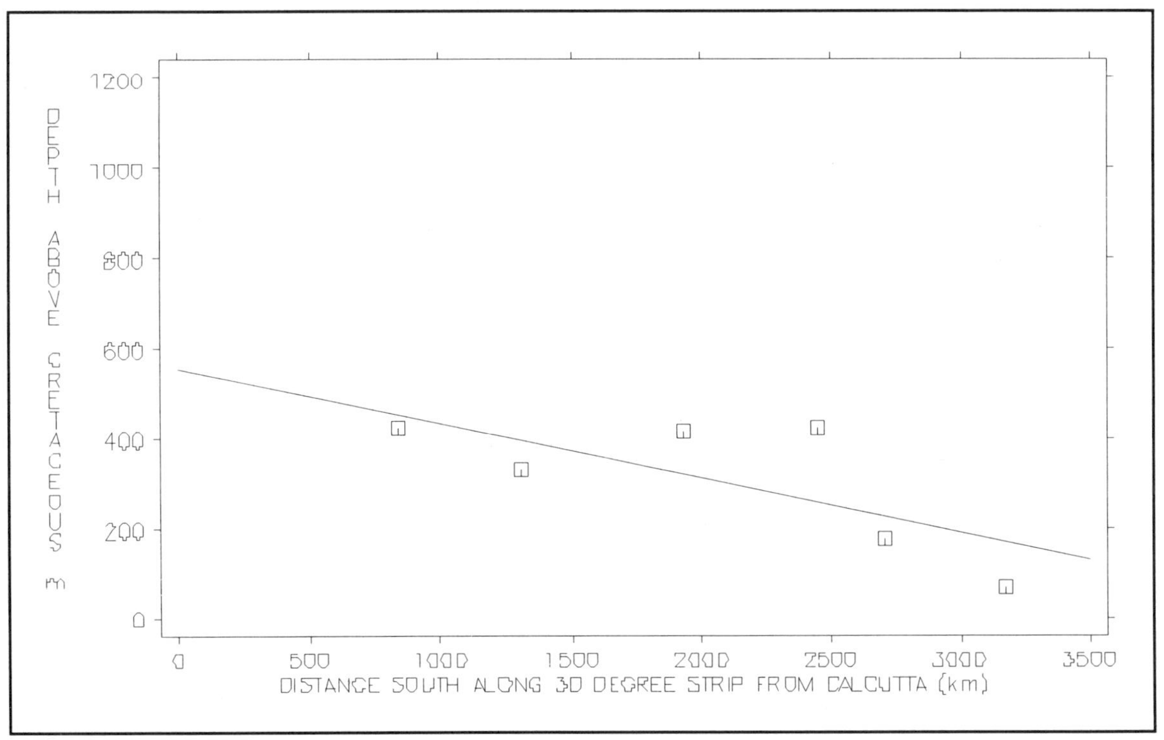

Fig. 3.2 *Thickness of sediment above the Cretaceous/Tertiary boundary in the Indian Ocean vs. distance southward from Calcutta.*

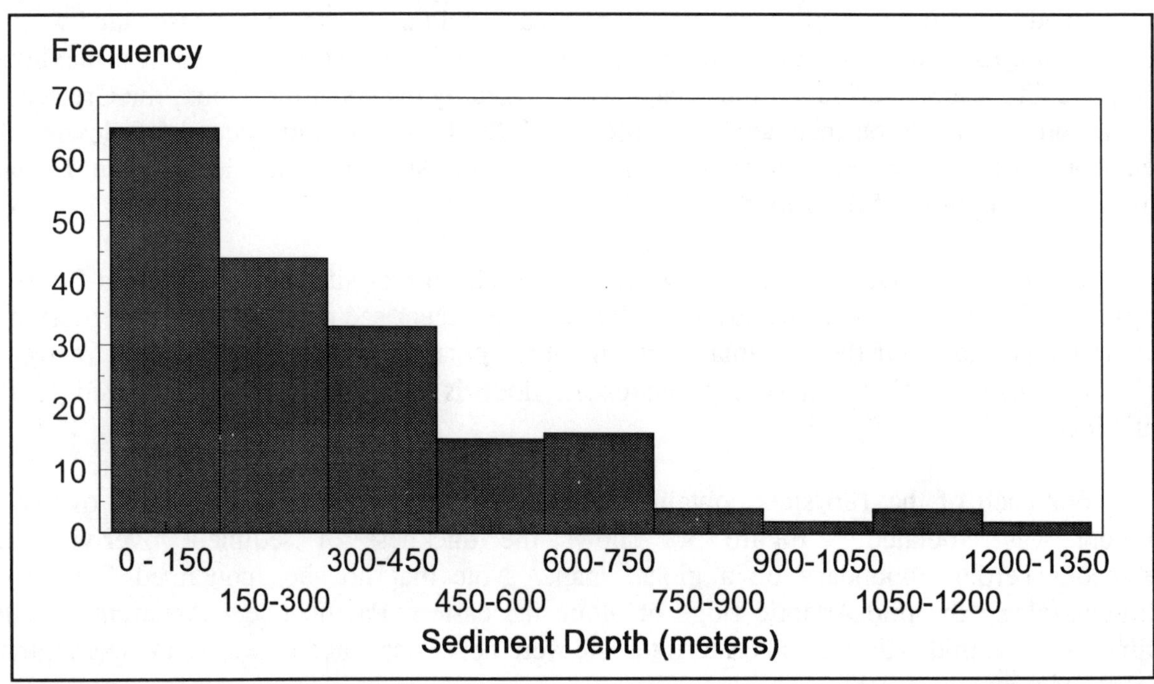

Fig. 3.3 *Frequency histogram of sediment thickness above the Cretaceous/Tertiary boundary for all DSDP cores.*

Sediments near the continents and the mouths of rivers appear to be somewhat thicker. Figure 3.2 shows a plot of thickness of the sediments southward from the mouth of the Ganges River at Calcutta. Note the decrease in thickness as a function of distance from the mouth.

Sediments above the Cretaceous appear to be thickest in the central Pacific. The mean thickness of the sediments above the Cretaceous/Tertiary boundary for all 186 sites was 322 meters with a standard deviation of 273 meters. Figure 3.3 shows a histogram of sediment depth for the 186 sites.

A Young-earth Age Model

The conventional age model used to calculate the age of sediment as a function of depth assumes that the accumulation rate of sediment was essentially constant over millions of years at today's rate of about 2×10^{-5} meters/year. If, in fact, the accumulation rate was much greater following the Flood and decreased exponentially until today, then the period of time back to the formation of a given layer can be found by the following sediment accumulation model.

Let the sediment accumulation rate be an exponentially decreasing function of time since the Flood:

$$\frac{dy}{dt} = A\,e^{-\frac{t}{\tau}} \qquad 3.1$$

where y represents the height of a sediment layer above a reference point (in this case the Cretaceous/Tertiary boundary), A is a constant to be determined from the boundary conditions, τ the relaxation time, and t the time after the Flood when a layer of sediment was laid down.

This equation can be integrated to give the height y directly:

$$y = -A\tau\, e^{-\frac{t}{\tau}} + C \qquad 3.2$$

where C represents a constant of integration to be determined from the boundary conditions. For the first boundary condition, $y = 0$ at $t = 0$. No sediment had yet begun to accumulate, so:

$$y = 0 = -A\tau + C \qquad 3.3$$

Solving for C and substituting into Eq. 3.2:

$$y = A\tau[1 - e^{-\frac{t}{\tau}}] \qquad 3.4$$

For the second boundary condition, $y = H$ at $t = t_F$, where H represents the total depth of the sediment above the Cretaceous/Tertiary boundary and t_F is the time in years since the Flood. For this condition:

$$y(t = t_F) = H = A\tau[1 - e^{-\frac{t_F}{\tau}}] \tag{3.5}$$

Solving for A:

$$A = \frac{H}{\tau[1 - e^{-\frac{t_F}{\tau}}]} \tag{3.6}$$

Substituting back into Eq. 3.4:

$$y = \frac{H[e^{\frac{t}{\tau}} - 1]}{[e^{\frac{t}{\tau}} - \frac{e^{-\frac{t_F}{\tau}}}{e^{-\frac{t}{\tau}}}]} \tag{3.7}$$

A more useful relationship may be found by inverting this equation to find t as a function of y, H, and τ.

$$t = -\tau \ln[1 - \frac{y}{H}(1 - e^{-\frac{t_F}{\tau}})] \tag{3.8}$$

This relationship is typically called an age model and is used to find the age of a layer based on its vertical position. At this point, it is not specific to any particular worldview and can be applied to any chronology by substituting any time frame, t_F, between the Cretaceous/Tertiary boundary and today.

If the chronology of the Biblical events according to Ussher (1786) is assumed to be true, approximately 4,500 years have transpired since the Flood ($t_F = 4{,}500$). Using this time interval, the average observed depth of sea-floor sediment above the Cretaceous/Tertiary boundary (322 meters), and the measured accumulation rate of sediment today, the relaxation time, τ, may be determined from Eqs. 3.1 and 3.5.

Substituting the time interval since the Flood and today's sediment accumulation rate into Eq. 3.1:

$$\frac{dy}{dt}(t = 4500) = Ae^{-\frac{4500}{\tau}} = 2 \ x \ 10^{-5} \ m/yr \qquad \textbf{3.9}$$

The initial sedimentation rate, A, in terms of the relaxation time, τ, may be found:

$$A = 2 \ x \ 10^{-5} m/yr \ e^{\frac{4500}{\tau}} \qquad \textbf{3.10}$$

Substituting A into Eq. 3.5:

$$H = 2 \ x \ 10^{-5} \ m/yr \ \tau[e^{\frac{4500}{\tau}} - 1] \qquad \textbf{3.11}$$

Rewriting in order to facilitate solving for τ:

$$\ln[1 + \frac{H}{2 \ x \ 10^{-5}\tau}] = \frac{4500}{\tau} \qquad \textbf{3.12}$$

This is a transcendental equation in τ. The solution for τ can be found using iterative methods or by finding the point at which the two sides of the equation are satisfied jointly. The second method was used here by plotting the left and right sides of Eq. 3.12 simultaneously and solving for τ using the average value of H. The solution to this transcendental equation gives a value for τ of 373 years.

Substituting τ = 373 years and t_F = 4,500 years into Eq. 3.8 results in the following young-earth age model, derived from young-earth boundary conditions:

$$t = -[373 years] \ln[1 - \frac{y}{H}(1 - e^{-\frac{4500 years}{373 years}})] \qquad 3.13$$

This age model is displayed in Fig. 3.4. The height of sea-floor sediment above the Cretaceous/Tertiary boundary, y, is shown on the horizontal axis and time since the Flood, t, on the vertical axis. The age model is shown for several total sediment depths, H. Note, that each curve assymtotically approaches the value of H as time approaches 4,500 years after the Flood. In general, it can be seen from Eq. 3.7 that $y = 0$ when $t = 0$ and $y = H$ when $t = t_F$.

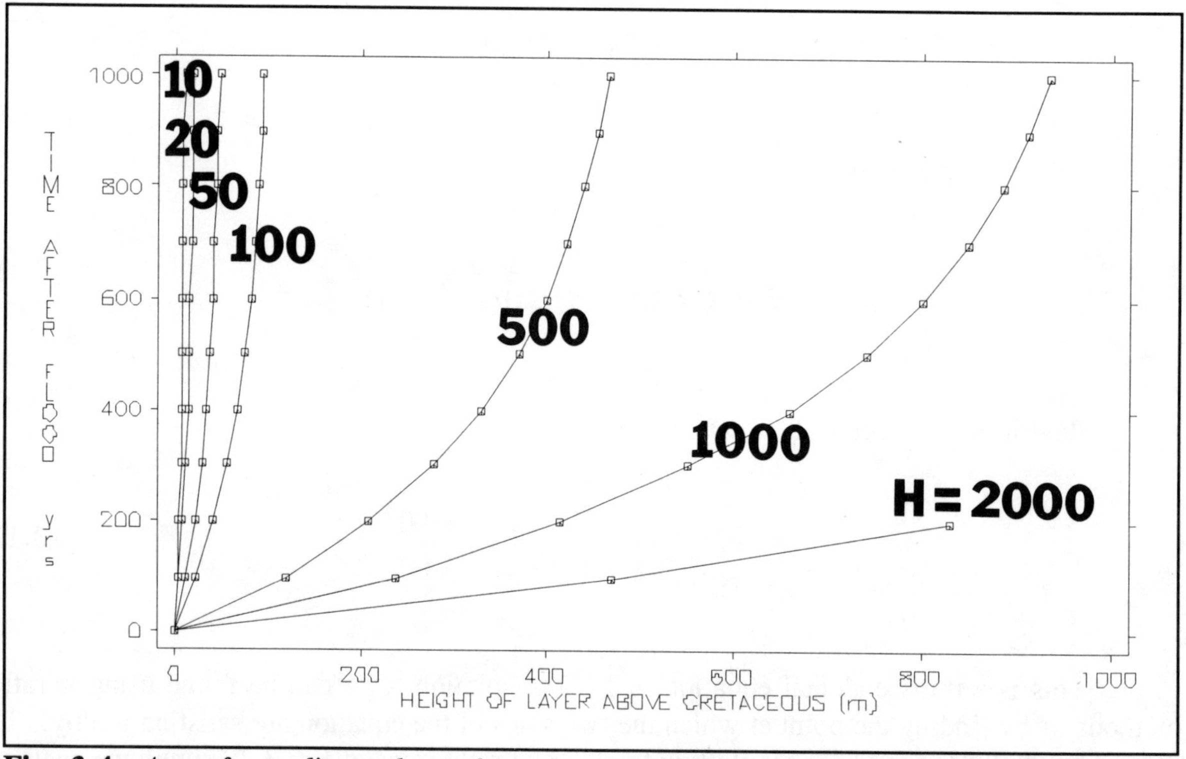

Fig. 3.4 *Age of a sediment layer from the young-earth age model vs. height above the Cretaceous/Tertiary boundary and the total sediment depth, H, in meters.*

Applications of a young-earth age model

The age model developed here can now be applied to data used by Kennett et al. (1977) to estimate ocean temperatures from the Cretaceous to the present. Documentation of the operations, lithology, and biostratigraphy for sites 277, 279, and 281 in the Ocean Drilling Project are summarized in Appendix E of this monograph. The analytical procedures and interpretations are contained in Shackleton and Kennett (1975). For this analysis the total sediment depth H above the Cretaceous/Tertiary boundary was found to be 760 meters. Fig. 3.5 shows the results of applying the new young-earth age model to these same data.

A significantly different interpretation of the data from that of Kennett et al. (1977) results. First, the period over which the data occur is assumed to be about 2 thousand years, rather than 65 million years. Second, the temperature initially decreases rapidly, followed by a slower decrease. The decrease shown by Kennett et al. (1977) is basically linear with a few short-period departures implying a gradual cooling over a long period of time. The trend shown in Fig. 3.5 is typical of rapid cooling driven by a large temperature gradient. If the oceans were initially warm at the end of the Flood and were cooled to a new equilibrium temperature by radiation to space in the polar regions, this would be the type of cooling curve one would expect. The relaxation time appears to be about 1 thousand years.

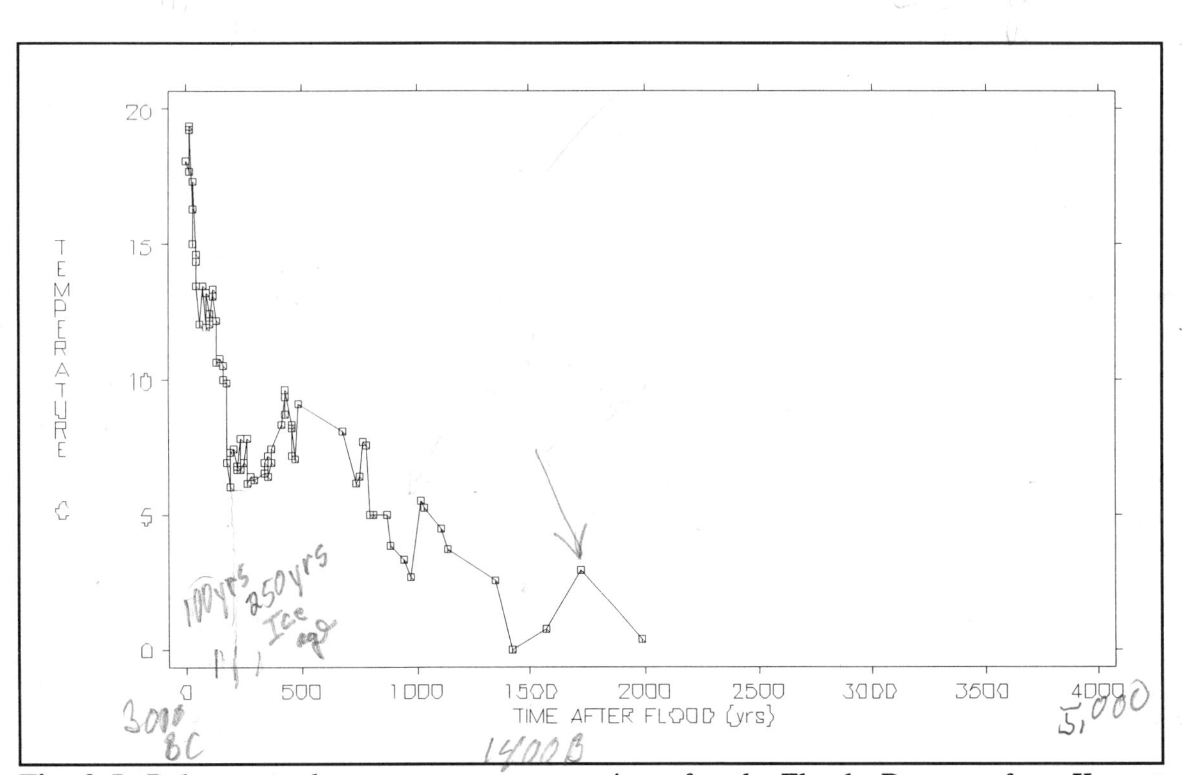

Fig. 3.5 *Polar ocean bottom temperature vs time after the Flood. Data are from Kennett et al. (1977) composited from DSDP sites 277, 279, and 281.*

This curve was derived from benthic forams in the South Pacific at high latitudes, so polar ocean bottom waters show dramatic cooling. Similar analysis of polar surface waters using planktic forams show a similar trend but average about 1 degree warmer. Equatorial surface waters show only a minor cooling of 5 degrees or so while equatorial bottom temperatures show a similar trend as polar waters.

These results are interpreted as surface cooling of polar waters followed by sinking and movement toward the equator along the ocean floor. A general oceanic circulation is established where warm equatorial water is transported poleward at the surface and cold polar water is transported toward the equator at the ocean floor. Horizontal gyres within the separate ocean basins are superimposed on these latitudinal motions by the Coriolis force.

In the polar regions one would expect surface cooling to decrease the temperatures at the ocean floor because the cooler water aloft would sink and displace the warmer water below. This interchange would result in vigorous vertical mixing and cooling of bottom waters. During this strong cooling period one would predict outstanding conditions for nutrient supply and formation of biogenous sediments in the polar regions. In the tropics the ocean would have become more stratified with time because of the advection of cold bottom water under the warmer surface water. Except for specific regions of upwelling along the continents and near the equatorial countercurrents, vertical transport of nutrients and, therefore, the formation of biogenous sediments, would have been more restricted.

The data resolution in Fig. 3.5 is very coarse. Near the top of the sediments sampling must occur at close intervals for the young-earth model because the sedimentation rate is decreasing exponentially. Fortunately, many cores have been extracted in recent years and sampled for $\delta^{18}O$ at very high resolution. This allows time to be resolved to short intervals near the top of the core.

Fig. 3.6 shows the results of applying the new young-earth age model to the data discussed in Appendix E. A high-resolution core was extracted from site RC11-120 in the Sub-Antarctic Pacific at about 45° S latitude. Note, that a consistent warming trend of about 5°C has occurred in the recent past preceded by rapid fluctuations at various time scales. Rapid warming followed by a slow cooling trend occurred between 1500 and 2500 years after the Flood. The "Ice Age" in the young-earth chronology would have ended about 2000 years ago. This event has been identified in the literature as the most recent "Ice Age" followed by rapid deglaciation. Note, that the period of this event is on the order of 700 years for the young-earth model instead of the conventional 100,000 years.

The young-earth age model has also been applied to a second high-resolution core also shown in Appendix E. A high-resolution core was taken from site V28-238 in the Pacific near the equator. The results, shown in Fig. 3.7, also show a 5 degree warming trend in the recent past preceded by similar oscillations in temperature. The period of the feature in this core associated with the most recent "Ice Age" is also about 700 years, but the temperature is about 15 degrees warmer. Because this core was longer than the previous one we can see a longer

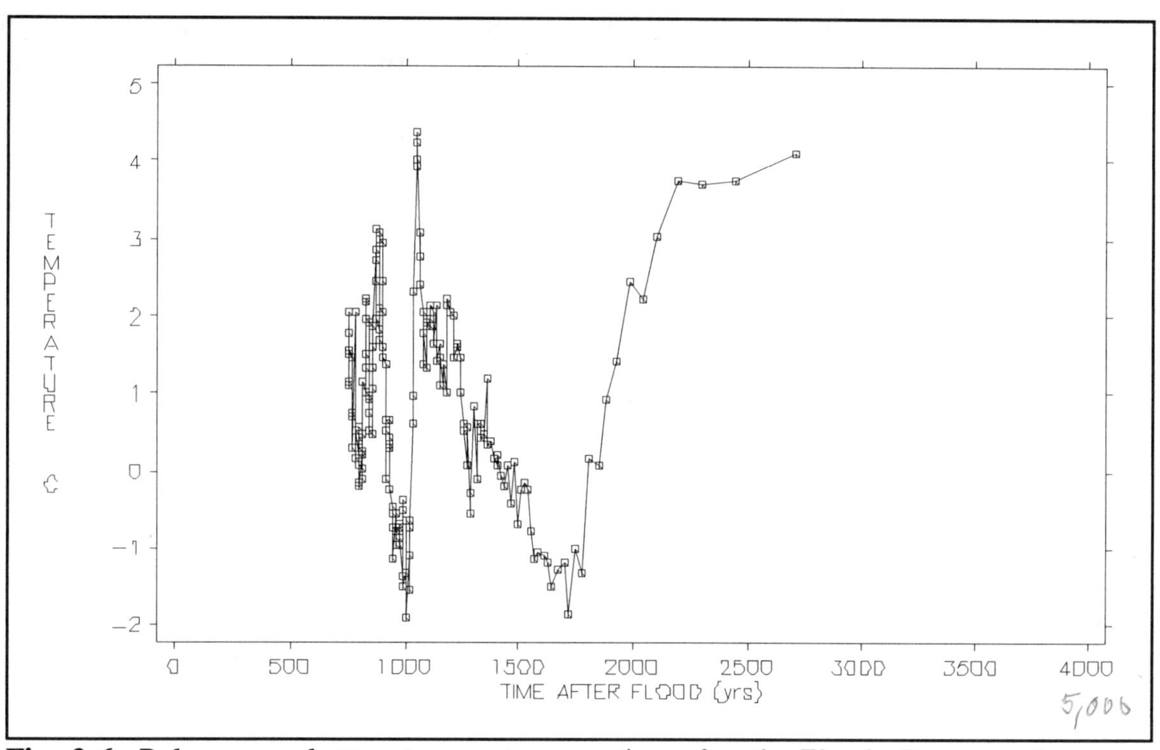

Fig. 3.6 *Polar ocean bottom temperature vs. time after the Flood. Data are from core RC11-120 used in the CLIMAP project.*

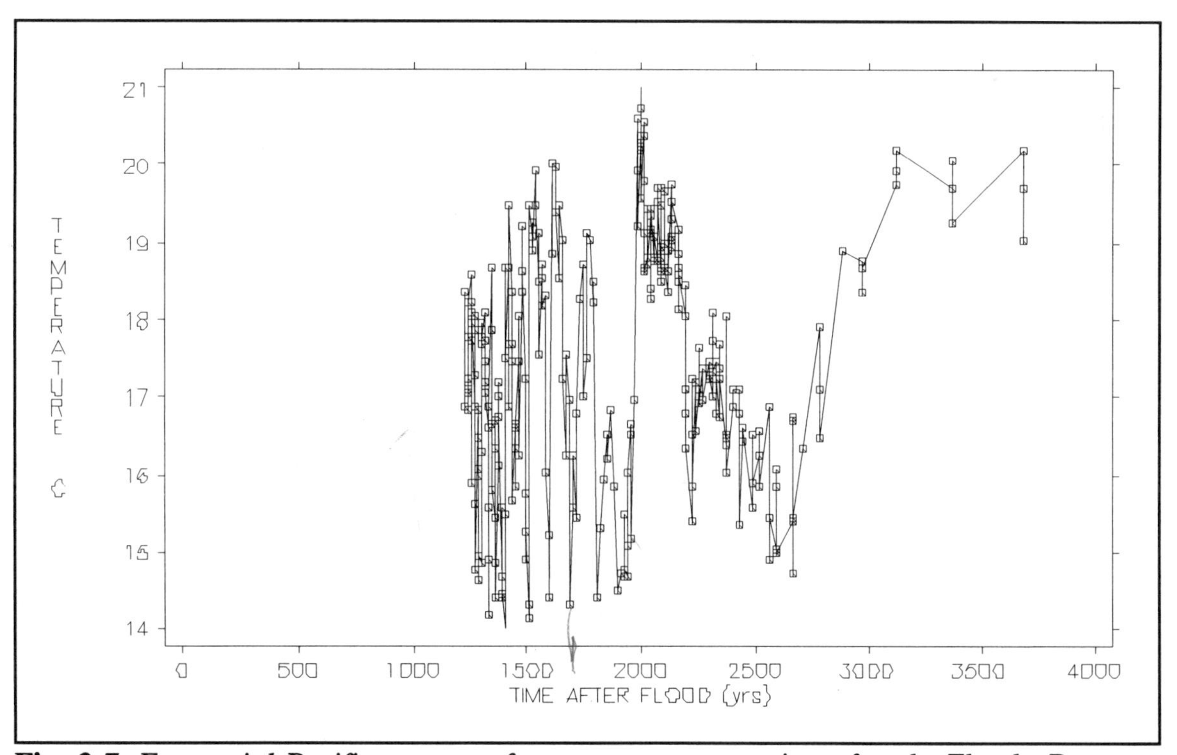

Fig. 3.7 *Equatorial Pacific ocean surface temperature vs. time after the Flood. Data are from core V28-238 used in the CLIMAP project.*

period of temperature oscillations into the past. Notice that these oscillations have a fairly uniform period of about 100 years. This compares to a period of about 20 thousand years derived from the conventional model.

Application of the young-earth age model to sediment cores results in major differences in the periods and timing of climatic events. If this alternative age model is correct, new interpretations of paleoclimatology will result.

CHAPTER 4

IMPLICATIONS OF A YOUNG-EARTH AGE MODEL

It has been recognized for several years that the layering of sediments on the ocean floor have been laid down in such a manner that some type of harmonic process has occurred. Analysis of $\delta^{18}O$ in fine resolution cores show periodic repetitions of cold and warm periods. A statistical correlation between the temperature oscillations and the periods of the three orbital parameters of the earth/sun system has led to stronger support for the Astronomical Theory. CLIMAP and SPECMAP were two projects designed to strengthen this relationship.

A frequency analysis of many cores with the traditional age model found that peaks in the frequency spectra occurred at periods of approximately 20, 40, and 100 thousand years. Because these periods were similar to those of the orbital parameters, it has been assumed that the driving mechanism for the temperature fluctuations derived from sea-floor sediments is the change in radiational warming of the earth as the earth/sun distance and orientation change. These concepts have become known as the Astronomical Theory, a revision of a theory proposed by Milankovich (1930, 1941).

However, several difficulties have yet to be resolved with this theory. First, the magnitude of the change in radiational heating calculated from the orbital parameters does not seem to be large enough to explain the observed cooling and heating. Secondary feedback mechanisms have been proposed to amplify the orbital effects. However, it has been found that many of the hypothetical feedback mechanisms are of the wrong sign at certain phases of the orbital cycles.

A major result of this need for feedback mechanisms has been the development of a perspective that the earth's climate systems are extremely sensitive to minor disturbances. A relatively minor perturbation could initiate a non-linear response which could lead to another "Ice Age" or "Greenhouse." Because of the fear of the consequences such a small perturbation might cause, radical environmental policies on the release of smoke, chemicals, and other pollutants and the cutting of trees have been imposed by some countries. If the basis for the Astronomical Theory is wrong, many of the more radical environmental efforts may be unjustified.

A second difficulty with the Astronomical Theory is the relative effect of the orbital parameters. The orbital parameter which has a period of about 100 thousand years and is associated with the 100 thousand-year-long "Ice Ages" produces the weakest change in radiational heating. If the "Ice Ages" are caused by radiational changes, the orbital parameter causing them should be the largest of the three. Yet, the orbital parameter with the 100 thousand year period is the smallest of the three.

If the young-earth age model proposed by this work is valid, the correlation between sea-floor sediments and the orbital parameters is completely false. The periods illustrated in Figs. 3.6 and 3.7 are on the order of 100 years and 700 years. Rather than an external forcing function like orbital parameters causing fluctuation in the earth's climate system, it is suggested that these oscillations are a manifestation of frequencies which are naturally present in the earth-atmosphere-ocean system. These natural frequencies were probably excited by the initial high-energy events of the Flood. In the young-earth model there has been only enough time for one "Ice Age" since the Flood. The initial forcing function for the "Ice Age" was the tremendous amount of heat left in the oceans by the events of the Flood. The length of the "Ice Age" would have been determined by the amount of time for the oceans to lose their heat to the atmosphere and subsequently to space.

Many other shorter-period oscillations in the earth's climate system may still be operating, however. For example, a significant oscillating climate event which has received a large amount of international research attention recently is the El Nino/Southern Oscillation (ENSO) which has been documented in the equatorial Pacific (Jacobs et al., 1994). This climate event starts as a warming of surface waters in the western equatorial Pacific. It progresses eastward over a period of 2-4 years increasing precipitation along the equator and changing the wind patterns. When it intersects the Americas, it produces flooding and major changes in marine habitats along the west coasts of both continents. Effects further east cause wet and dry regions over large areas. This oscillation has a period of about 7 years and may be just one example of many such oscillations still observable in our atmosphere/ocean system.

If a young-earth model of sea-floor sediment accumulation such as that developed in this monograph can be justified, the conventional theories of multiple "Ice Ages", Greenhouse Warming, and millions of years of earth history required for evolutionary processes will be refuted. There seems to be no middle ground between the young-earth and old-earth models when treating the accumulation of sea-floor sediments.

CHAPTER 5

CONCLUSIONS AND RECOMMENDATIONS

An alternative, analytic, young-earth model of sea-floor sediment accumulation has been developed in this treatment. Rather than a slow accumulation of sediments at a nearly constant rate of a few centimeters per millennium over millions of years, an initially rapid accumulation of sediments decreasing exponentially to today's rate over some 4,500 years was assumed. Observations of $\delta^{18}O$ from sea-floor sediment cores were transformed to estimates of temperature and plotted as a function of time of deposition in accordance with this exponential model.

These plots indicate that temperature at the floor of tropical and polar oceans and the surface of polar oceans decreased rapidly, immediately following the estimated end of the Flood. This decrease was on the order of 15°C and assymtotically cooled to today's average value of 4°C. The major portion of the cooling occurred in about 1000 years, in agreement with Oard's (1990) estimates of cooling following the Flood, resulting in the "Ice Age." Application of this model to very detailed tropical cores found a consistent warming trend of about 5°C over the recent past, preceded by rapid fluctuations of temperature at various time scales. The period of the longer fluctuations, typically identified with the "Ice Ages," is on the order of 700 years, rather than the conventional 100,000 years. The period of the shorter fluctuations is about 100 years, compared to the conventional 20,000 years.

The major decrease in oceanic temperature by 15°C, following the Cretaceous Period, is suggested to be the cooling of the ocean to a lower equilibrium temperature following the Genesis Flood. The 100-year and 700-year fluctuations are suggested to be transient oscillations as the ocean/atmosphere system reached equilibrium.

Massive quantities of data available from DSDP, ODP, and other sea-floor core drilling projects may be used to investigate other features of sediment accumulation from a young-earth perspective. $\delta^{18}O$ is only one of many variables available for such studies. Cores from almost 1000 sites and nearly every region of the ocean floor are available for study. It is likely that an entirely new understanding of paleoceanography could be developed from this preliminary age model.

In order to improve the young-earth model proposed here, similar analyses should be made of $\delta^{18}O$ measurements for many additional cores. The results of Douglas and Savin (1971, 1973, 1975), Savin et al. (1975), and Shackleton and Kennett (1975) should be replicated with more recent cores over a wider geographic distribution. $\delta^{18}O$ observations from the upper 50 meters of sediment would be of particular interest.

Further consideration should be given to the identification of the Flood/Post-Flood boundary. It may be that the Cretaceous/Tertiary boundary is too deep in the geologic column. A larger survey of sediments above the Cretaceous/Tertiary boundary may lead to smaller values for a typical thickness, reducing the model accumulation rate and revising other parameters in the young-earth model. A universal average sediment thickness should not be used to plot time versus depth at any single site. The constants in Eq. 3.8 are site specific.

An analysis of the productivity of biogenous sediments in a post-Flood ocean should be made and compared with the mass of sediments observed. The accumulation of hundreds of meters of sediment, on the average, and kilometers of sediment in some locations, such as the Arctic Ocean, require very high productivity following the Flood. Although the potential for high productivity has been suggested by Roth (1985), can the oceans supply enough nutrients, in some 4,500 years, to explain the observed sediments?

Refinements in the young-earth model should be made to better simulate the formation of sediments. Such assumptions as the exponential decrease in accumulation, the total depth of post-Flood sediments, and the composite of biogenous and lithogenous sediments should be explored further. The model may need separate parameters for different oceans, latitudes, and sediment types, as well as sites.

A similar study should be conducted for $\delta^{13}C$. $\delta^{18}O$ was selected for this first study because of its immediate relationship to climate and the polar ice sheets. However, the burial of carbon has major implications on the mass balance of carbon in the hydrosphere, biosphere, and atmosphere. It affects the formation of carbonates, the radiation balance and temperature of earth, and paleochronometers such as ^{14}C. Combinations of $\delta^{18}O$ and $\delta^{13}C$ may be useful for estimating productivity and sediment accumulation rates.

The result of this effort was to initiate the development of an analytical model of sea-floor sediment accumulation. The model uses the measured sediment accumulation rate of today, the observed sediment depth on the ocean floor, and a literal Biblical time frame as boundary conditions. An exponentially-decreasing accumulation function was assumed. All of the questions have not been answered. In fact, this monograph may raise more questions than it answers. Other researchers are encouraged to work on portions of this problem and keep this author informed.

APPENDIX A

SEA-FLOOR SEDIMENT

Types and Sources of Sea-floor Sediment

Except for relatively few areas of exposed rock, the ocean bottom is carpeted with sediment. These deposits consist of insoluble particles from the continents, mixed with the undissolved shells and skeletal remains of marine organisms. Using seismic techniques geophysicists have studied ocean bottom structures and sediment depths (See Fig. A .1). These data show that sediment thickness varies widely. In the Pacific Ocean, the uncompacted sediment is about 600 meters thick. In the Atlantic Ocean it varies between 500 and 1000 meters. In the Puerto Rico Trench, the sediment thickness exceeds 9 kilometers, and in parts of the Arctic Sea it exceeds 4 kilometers. The overall uncompacted thickness of sea-floor sediments in the deep ocean averages about 600 meters. Deep-sea sediment commonly contains 50% water by volume. Thus if we were to compact these sediments, squeezing out the water,

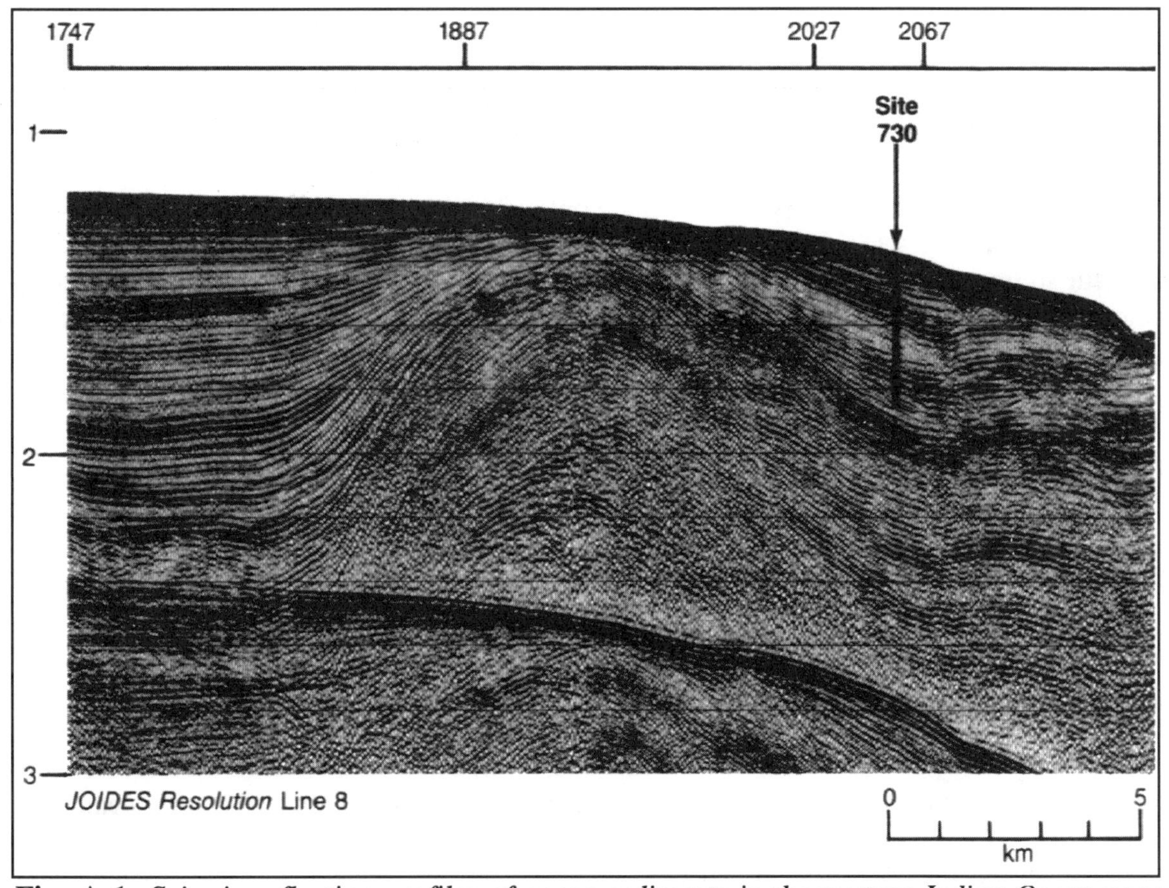

Fig. A.1 *Seismic reflection profiles of ocean sediments in the western Indian Ocean near Arabia. The vertical scale is highly exaggerated and numbers are approximately 100 meters of depth.*

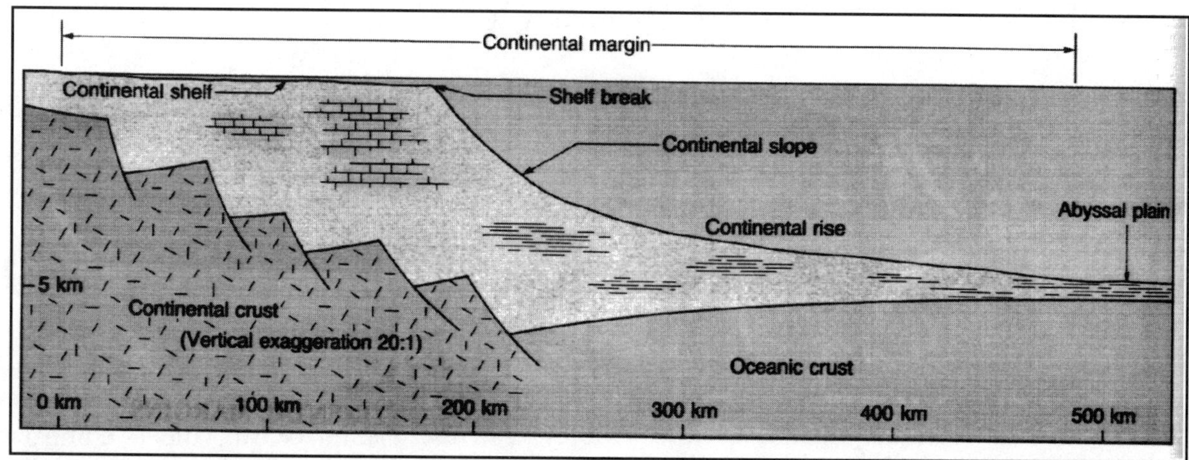

Fig. A.2 *Schematic profile of the continental margin. Notice that the continental shelf is underlain by much thicker sediments than the midocean.*

the compacted sediment would form a layer only 300 meters thick.

Sediment thickness tends to be greater near continents, particularly near the mouths of major rivers. Near continents the type of sediment is generally more sandy, and in deep oceans more muddy. Sand is defined as sediment made of grains larger than 62 microns in diameter. Mud is defined as sediment made of grains smaller than 62 microns in diameter.

In general, sands are deposited on the continental shelf and commonly occur along the shore, forming beaches (See Fig. A.2 for a schematic of the continental margin). Sands may also occur in irregular patches at greater depths on continental shelves. These irregular patches appear to be the result of now-inactive processes. Such sediments are known as relict sediments because they were not deposited by processes now active.

Mud is the most common sediment covering the deep-ocean bottom as well as the continental shelf and slope. In the deep-ocean, sand-sized particles generally constitute less than 10% of the sediment. In the North Atlantic Ocean, approximately 20% of the deep-sea sediment samples contain sand layers, thought to have been transported there by currents flowing down the continental slopes.

Classification of Sediment Origin

Lithogenous (derived from rocks) - components are primarily mineral grains derived from rocks and soils formed on the continents and transported to the ocean by rivers or wind. Volcanoes on land or under the sea also contribute lithogenous particles such as volcanic ash.

Biogenous (derived from organisms) - components are skeletal remains, such as shells, bones, and teeth of marine organisms. They contribute calcium carbonate, silica (opal), or phosphate minerals to the sediment.

Hydrogenous (derived from water) - components originate from inorganic chemical reactions in seawater or in the sediment itself. An example is manganese nodules.

Lithogenous Sediment

Lithogenous sediment consists primarily of mineral grains, derived from the breakdown of rocks during soil formation on the continents. These sediments are carried by running water into rivers by erosion and on down to the sea, small particles being suspended in the water and larger ones dragged along the river bottom by the energy of the flowing water.

The particles are deposited when the water movements are too weak to carry the grains farther. On the continental shelf, wave action or currents sort sediment particles by size, depositing the large particles and removing the smallest. The largest particles tend to remain near the shore, where they may form beaches. Smaller particles are commonly transported seaward and deposited in deeper water on the continental shelf and slope. These tiny particles may also be carried by the ocean surface currents or moved by near-bottom currents. The resulting fine-grained sediments, usually muds, are frequently deposited in bands generally parallel to the coastline.

Despite the movement of the sediment particles after reaching the ocean, most do not travel far. A large fraction of the sediment brought to the ocean by a river tends to be deposited near its mouth. Indeed, it appears that most lithogenous sediment does not travel beyond the continental shelf and slope. Most sand is hydraulically trapped at the shoreline — even during large storms.

The smallest particles may be carried away from the continents by water movements, because minute particles take several years to settle to the ocean bottom. Particles less than 0.5 microns in diameter take several hundred years to pass through the oceans and reach the bottom. During this time the particles are carried thousands of kilometers by ocean currents. Some fine particles are also carried by winds. Hence virtually every part of the ocean receives some lithogenous sediment. Far from continents, lithogenous sediment accumulates slowly today, forming a layer approximately 1 mm thick in a thousand to ten thousand years (see chapter 3). Near the continents, and especially near large rivers such as the Amazon or Congo Rivers, lithogenous sediment accumulates much more rapidly.

The long transit time of very small particles in the ocean not only results in their wide dispersal, but also provides ample opportunity for chemical reactions to take place. For example, iron in the water or on the particle can react with dissolved oxygen, forming a rust-like iron oxide coating on mineral grains. The abundance of such red or brown-stained grains in deep-sea sediment accounts for their colors and their names, red-clay and brown-clay. Colors of deep-sea sediments range from brick red in the Atlantic to chocolate brown in the Pacific Oceans.

In contrast, rapidly accumulating sediment on the continental shelf and slope are in contact with seawater for too short a period to react fully with dissolved oxygen in the water. Thus these grains do not commonly acquire a reddish or brownish color. Such sediments have a variety of colors and are often referred to as green muds or blue muds. Green and blue colors are especially conspicuous in sediments accumulating in areas overlain by ocean waters having unusually low dissolved-oxygen contents. Lithogenous particles transported by winds are especially important in the two east-west belts centered around 30°N and 30°S. High mountains and the world's deserts are prime sources for air-borne dust particles. Where the influx of lithogenous grains from rivers is low, air-borne particles are an important source of sediment.

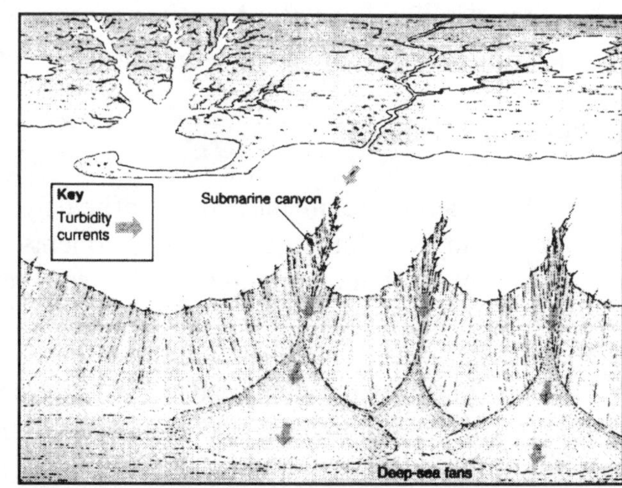

Fig. A.3 *Turbidity currents move downslope cutting submarine canyons into the continental margin and depositing layers of sediment as deep-sea fans.*

Fine-grained volcanic ash blown, into the atmosphere during volcanic eruptions, settles out on the adjacent ocean surface. Volcanic ash particles are incorporated in the deep-sea sediment accumulating in volcanic regions such as the Indonesian or Aleutian areas. The mineral grains and volcanic rock fragments are often easily recognized in the sediments. Under favorable conditions, volcanic fragments or layers of ash from known large volcanic eruptions can be identified in the surface sediments.

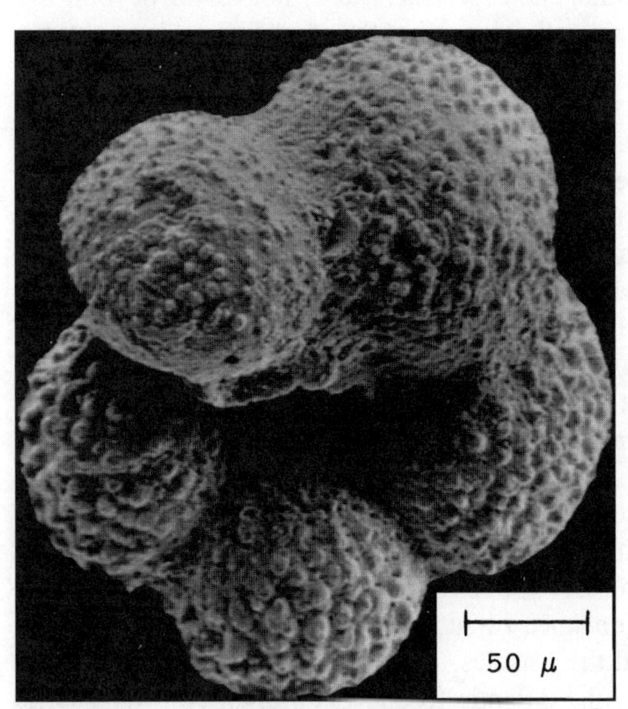

Fig. A.4 *A Planktic foram common in the Cretaceous sediments of Maud Rise of the Weddell Sea near the southern tip of South America.*

Of the estimated 10,000 volcanoes on the ocean bottom, only a few, such as the Hawaiian Islands in the Pacific or the Azores in the Atlantic, penetrate the ocean surface forming islands. It is likely that eruptions of submarine volcanoes contribute large amounts of lithogenous sediment to the ocean floor. Because of the difficulty of observing these volcanic eruptions, our knowledge of their sediment contribution remains limited.

Glaciers are another source of lithogenous sediments. As glaciers flow from the interior to the coast, underlying rock surfaces are worn down by the removal of fine rock fragments. In addition, the ice picks up and carries rocks and boulders of various sizes. When the glacier flows into the ocean, blocks of ice break off, forming icebergs. As they melt, the icebergs release their sediment load. The resulting ice-rafted (glacial-marine) sediment — a mixture of mud, sand, and boulders — covers much of the Antarctic continental shelf and some of the North Pacific deep-sea floor. Much of the North Atlantic has glacial-marine sediments, much of it due to sea ice and debris flows.

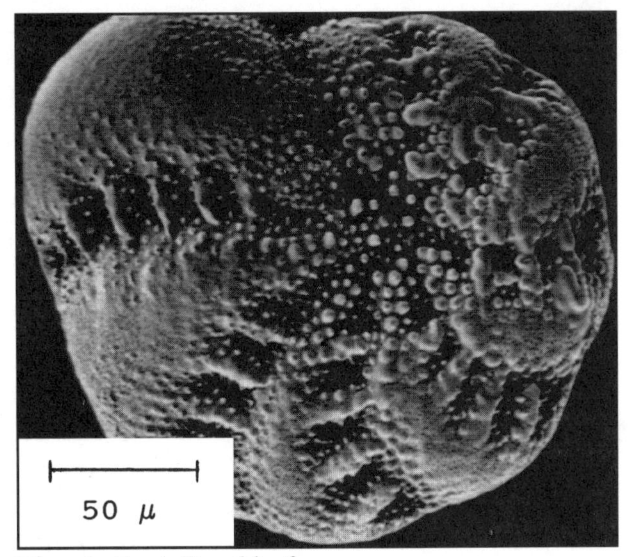

Fig. A.5 *A Benthic foram.*

During the "Ice Age," large parts of the Northern Hemisphere continents were covered by continental ice sheets, much as Antarctica is presently, and the adjacent continental shelves were covered by ice-rafted sediment. Continental shelf areas off formerly glaciated coasts often do not receive sufficient modern sediment to bury these relict sediments, leaving them exposed at the surface, e.g., on the continental shelf off the New England coast.

Fig. A.6 *A coccolith from Maud Rise in the Weddell Sea near the southern tip of South America.*

Before 1950, most oceanographers and geologists believed that deep-sea deposits accumulated only by individual grains settling slowly through the water. Investigations of submarine telegraph cable breaks associated with earthquakes in several parts of the world indicate that currents of relatively dense, mud-rich water, called turbidity currents, flow down the continental slope and onto the ocean floor. Such flows, actually observed in Lake Mead, on the Colorado River, and in certain European lakes, transport to the deep-ocean bottom sandy sediment originally deposited on the continental shelf or slope.

Submarine cable breaks, presumably caused by turbidity currents (See Fig. A.3), are common near major rivers, such as the Congo River, which carry large amounts of sediment over large areas of the ocean floor. A turbidity current in 1929, caused by an earthquake on the Grand Banks south of Newfoundland, is thought to have deposited sediment on the deep ocean floor over an area about 1,000 kilometers long and 100 to 300 kilometers wide. Data on the time elapsed between the earthquakes and the breaking of submarine cables indicate that some turbidity currents travel at speeds of 20 kilometers per hour and faster. Turbidity currents are also fairly common along the continental slope. Debris flows have been observed from the Amazon fan down into the deep sea (Damuth and Embley, 1981).

Although large turbidity currents have not been directly observed in the deep ocean, they appear to be important sediment-transporting agencies. Turbidity currents and debris flows explain the common occurrence of deep-sea turbidites -- displaced sands with unusual textures, containing abundant shells and other remains of shallow water organisms. Such currents appear to be a major factor in forming and removing the sediment deposited in submarine canyons. Submarine canyons near the mouth of large, sediment-laden rivers, such as the Congo, would soon fill up if they were not occasionally emptied.

Sediment-rich flows are much more dense than normal seawater, causing the mixture to sink and flow along the ocean bottom. It appears that turbidity currents often flow along channels on the ocean bottom, much as a large river does on land. Obstructions on the bottom deflect these currents and even prevent their escape onto the deep ocean floor. This is especially obvious in the Pacific, where island-arcs and trenches obstruct the flow of water toward the adjacent ocean bottom. In the Atlantic, Indian, and Arctic oceans, turbidity currents can flow over and cover pre-existing topography with thick blankets of sediment and form large, rather featureless plains. Tops of seamounts, submarine ridges, and banks are not usually affected by turbidity currents. There, sediment accumulates by the particle-by-particle deposition of individual grains, and the deposits appear to be draped over the pre-existing topography rather than burying it completely.

Biogenous Sediment

Skeletal remains of organisms are important constituents of certain deep-sea sediments. Biogenous constituents may be divided into three main groups, based primarily on their chemical composition and secondarily on the organisms from which they are derived. In the order of abundance they are:

Calcareous constituents - primarily calcium carbonate, are derived from shells of a group of one-celled animals called Forams (See Fig. A.4 and A.5), platelets secreted by a tiny, one-celled algae, called Coccoliths (See Fig. A.6), and, in parts of the Atlantic, shells of tiny, floating snails, called Pteropods.

Siliceous constituents - primarily opal, the hydrated silica ($SiO_2 \bullet H_2O$) derived principally from shells of one-celled algae, called Diatoms (See Fig. A.7) and one-celled animals called Radiolaria (See Fig. A.8).

Phosphatic constituents (rich in phosphate) - primarily derived from bones, teeth, and scales of fish and other marine vertebrates. Areas such as offshore banks where fish are abundant may be covered by sediment composed in large part of such skeletal debris.

Any deposit containing more than 30% biogenous constituents by volume is known as an ooze. Since most deep-sea sediments are muds, the terms calcareous muds and siliceous muds are frequently used. Muds rich in phosphate are rare in the deep sea, although they are sometimes found covering submarine ridges or banks. Calcareous muds cover nearly half of the deep-ocean bottom and are most abundant in the shallower parts (less than 4,500 meters). They accumulate at rates today of between one and four centimeters/1000 years.

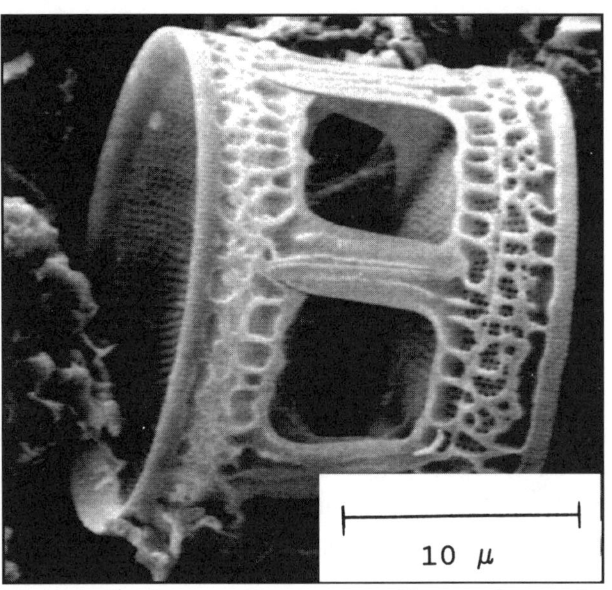

Fig. A.7 *Lower Cretaceous diatom from site 693 in the Weddel Sea near the southern tip of South America.*

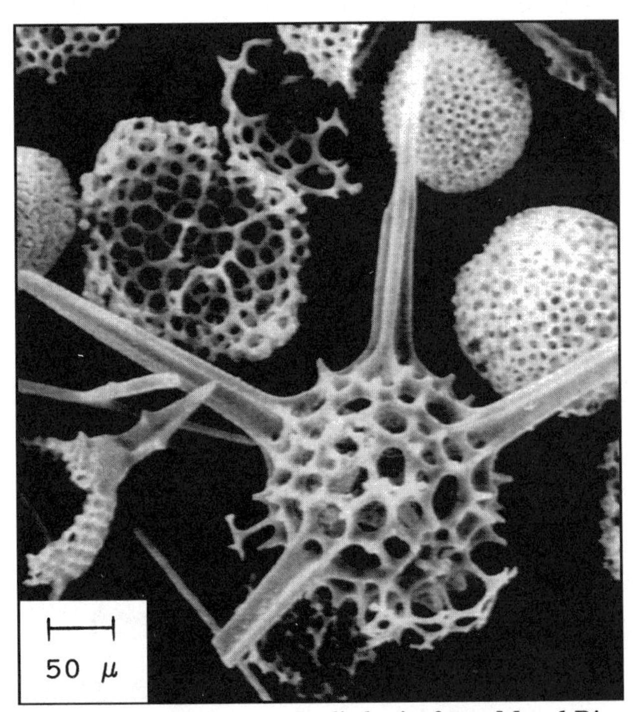

Fig. A.8 *Cretaceous radiolaria from Maud Rise in the Weddell Sea near the southern tip of South America.*

Siliceous muds occur principally in the Pacific Ocean, where their presence is not masked by large amounts of lithogenous sediment. Diatom-rich muds nearly surround Antarctica and occur in the North Pacific Ocean; radiolaria-rich sediment occurs along an east-west equatorial belt from Indonesia to Central America in the Pacific Ocean.

In the earlier discussion of lithogenous sediment only the source and transportation of the sediment particles was considered. When biogenous sediment is discussed, however, not only the source -- generally production by organisms living in the sunlit waters near the ocean surface -- but also destruction of the biogenous particles must be considered. Marine organisms remove the materials they require for skeleton formation from seawater (silicate, calcium, and phosphorus). Since the nutrients required for production are continually removed from the seawater and deposited on the ocean bottom, the nutrients remaining limit the concentration of many of these groups in the photic zone near the surface.

Most large organisms are eventually eaten and their skeletons or shells broken into smaller pieces. As these pieces sink below the calcium carbonate compensation depth, the more soluble shells tend to disappear. For calcareous shells, dissolution is more rapid in deeper ocean waters and on the deeper ocean floor. Siliceous and phosphatic grains apparently dissolve at all depths. The smaller the grains and the longer they are in contact with ocean water, the more they dissolve. This occurs as the fragments sink through the water and after they fall to the bottom but remain unburied by later sediment from the overlying seawater.

Certain organisms live in profusion near the ocean surface while their skeletons are rarely, if ever, observed in deep-sea sediment. Thus a potentially important source of sediment may be partially or completely obscured by destruction of the biogenous constituents before or immediately after deposition.

Another factor is dilution. Since biogenous oozes must contain more than 30% biogenous constituents by definition, it is evident that if lithogenous particles are supplied in large amounts, biogenous material will not be sufficiently abundant to form an ooze.

Hydrogenous Sediment

Manganese nodules (See Fig. A.9) commonly occur at the sediment surface over much of the ocean floor. They appear as black, potato-sized nodules or sometimes larger slabs composed of a complex mixture of iron-manganese minerals. Photographs of the deep-sea floor indicate that they cover 20 to 50% of the Pacific Ocean bottom.

The manganese nodules form extremely slowly today, a few hundredths of a millimeter to a millimeter in a thousand years. Viewed on the atomic level, this corresponds to between one and one hundred atomic layers per day, one of the slowest chemical reactions known, if the oceans have been undisturbed for several hundred million years. There are some researchers who believe volcanism and biologic processes in the deep sea can form these nodules more rapidly.

Fig. A.9 *Manganese nodules at a depth of 5,323 meters on the floor of the Pacific Ocean near Tahiti.*

Because of the apparent exceedingly slow rate of accumulation, hydrogenous sediment is believed to be abundant only in ocean areas remote from major rivers and where biogenous constituents are scarce. In other areas, lithogenous or biogenous constituents dominate the sediment. For example, in the central Pacific Ocean, the belt of hydrogenous sediment is interrupted by radiolarian-rich siliceous muds near the equator. This does not mean that hydrogenous constituents do not form in areas of rapidly accumulating sediments. Instead of forming large nodules or slabs at the surface, small, pea-sized or smaller, micronodules of manganese and iron ore are embedded in the more abundant constituents. The old-earth model may have a problem here, because the large majority of manganese nodules occur at the surface of sediments. If the nodules grow so slowly for such a long time, they would soon be buried, unless there is a mechanism to keep them on the surface, such as bioturbation.

Distribution of deep-sea sediments

Fig. A.10 shows a map of deep-sea sediments over the floor of the world's oceans according to Barron (1981). Carbonate sediments, red clay, and siliceous/red clay are seen to cover major portions of the ocean floor. Siliceous sediments cover bands near the equator in the Pacific and Indian Oceans and along the west coasts of North and South America. Ice-rafted sediments are restricted to polar regions. Terrigenous sediment, which is deposited near the mouths of rivers, is shown to occur along the edges of continents.

Table A.1 shows the percent coverage of the deep-ocean basins by the type of mud. Table A.2 shows the percent constituency of the various types of mud by source. Data in Tables A.1 and A.2 are from Revelle (1944).

Table A.1 Distribution of Deep-Sea Sediment

Type of Mud	Coverage of Deep Basins (%)
Brown	38
Calcareous	48
Siliceous	14
Total	100

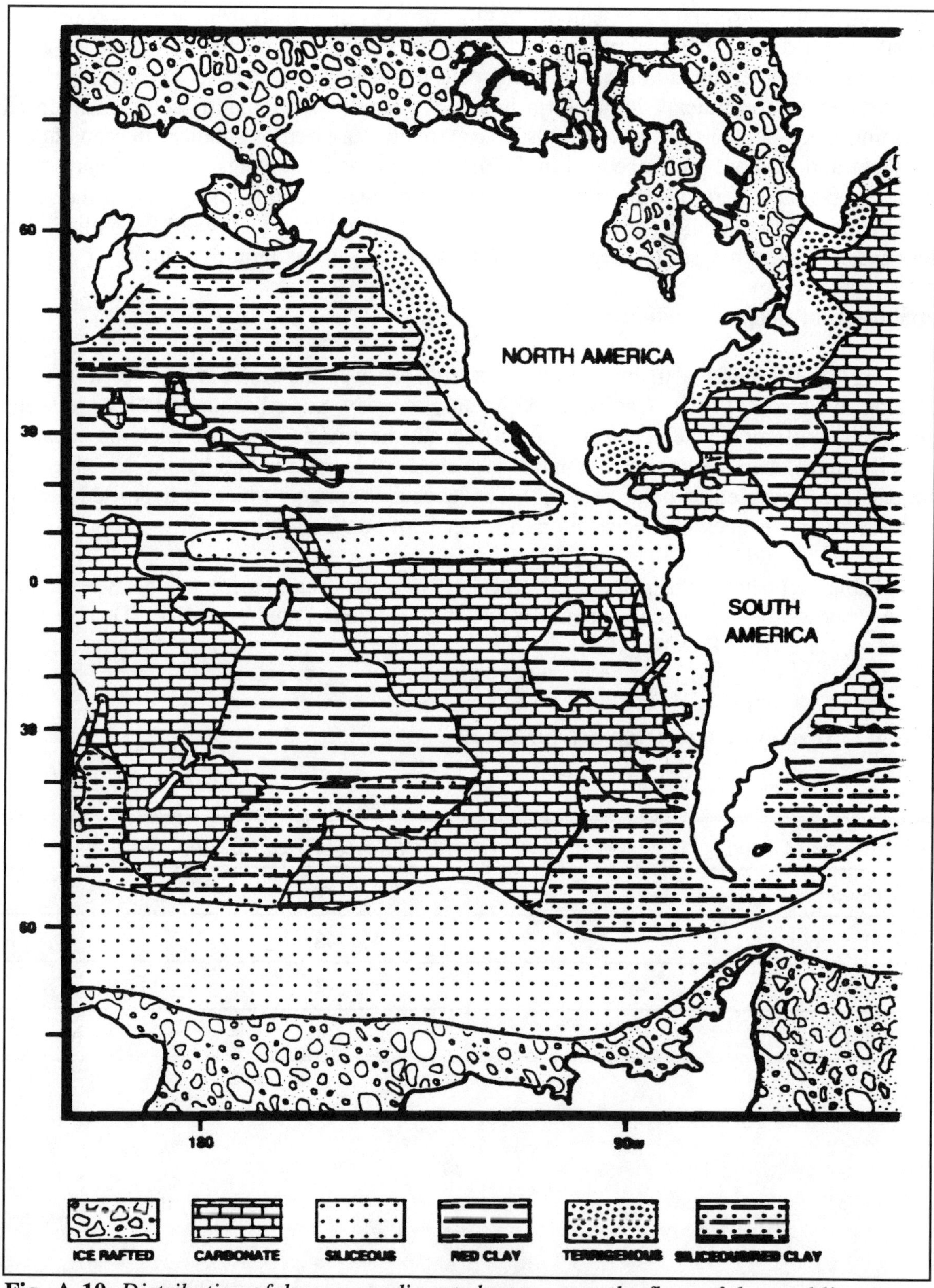

Fig. A.10 *Distribution of deep-sea sediments by type over the floor of the world's oceans.*

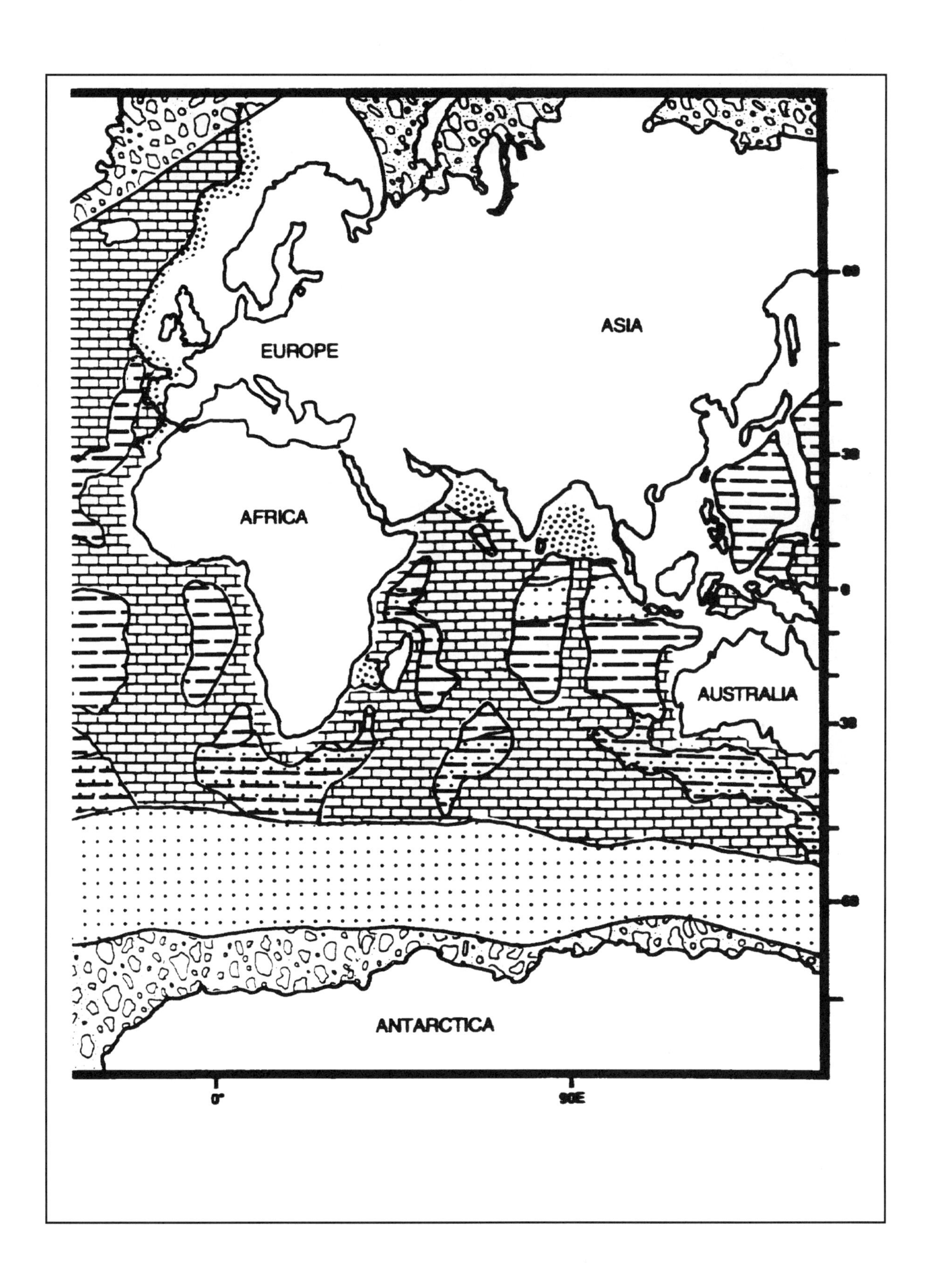
EUROPE
ASIA
AFRICA
AUSTRALIA
ANTARCTICA
60
30
0
30
60
0°
90E

Table A.2 Composition of Deep-Sea Sediments

Constituent	Brown Mud (%)	Calcareous Mud (%)	Diatomaceous Mud (%)	Radiolar-ian Mud (%)
Biogenous Calcareous	8	65	7	4
Biogenous Siliceous	1	2	70	54
Lithogenous & Hydrogenous	91	33	23	42
Total	100	100	100	100

APPENDIX B

DRILLING HOLES IN THE SEA FLOOR

Introduction

Cores have been extracted from the ocean floor since the 1960's over all the earth by ships specially equipped to operate in deep water. Two major projects, funded and directed by a multi-national consortium of research organizations, have operated in the oceans of the world. The Deep Sea Drilling Project (**DSDP**) operated from 1968 to 1983 using the drilling ship *Glomar Challenger*. Since 1983 the Ocean Drilling Project (**ODP**) has used the drilling ship ***JOIDES*** *Resolution*. Figure B.1 shows the ***JOIDES*** *Resolution* in port at San Diego in 1992. Prior to and during these two projects specially-equipped ships from individual research organizations also conducted independent additional drilling operations.

Fig. B.1 *The JOIDES Resolution in port in San Diego, California in December, 1992.*

Core samples have been extracted from over 600 sites by the **DSDP** and over 250 sites by the **ODP** as of 1992. Figure B.2 shows the sites drilled by **DSDP** and Figure B.3 those drilled by **ODP** through 1992. A complete list of drilling sites is itemized in Tables B.1 and B.2. Cores are typically drilled in 30 foot sections to depths of over 3,000 feet in the ocean floor. The core samples - cylinders of sea-floor rock and sediments about 3 inches in diameter and up to 30 feet long - have been analyzed by shipboard instruments and archived at the West Coast repository with the Scripps Institute at the University of California at San Diego, the Gulf Coast repository at the Texas A&M University Research Park at College Station, and the East Coast repository with the Lamont-Doherty Earth Observatory at Columbia University in New York.

Large volumes of on-board scientific results and later research results have been published by the consortium. Much of the data obtained by on-board analysis is available on CD-ROMs and floppy diskette from the National Geophysical Data Center in Boulder, Colorado. Data from special post analysis such as $\delta^{18}O$ are not available in this form but can be transcribed from the Research Results reports.

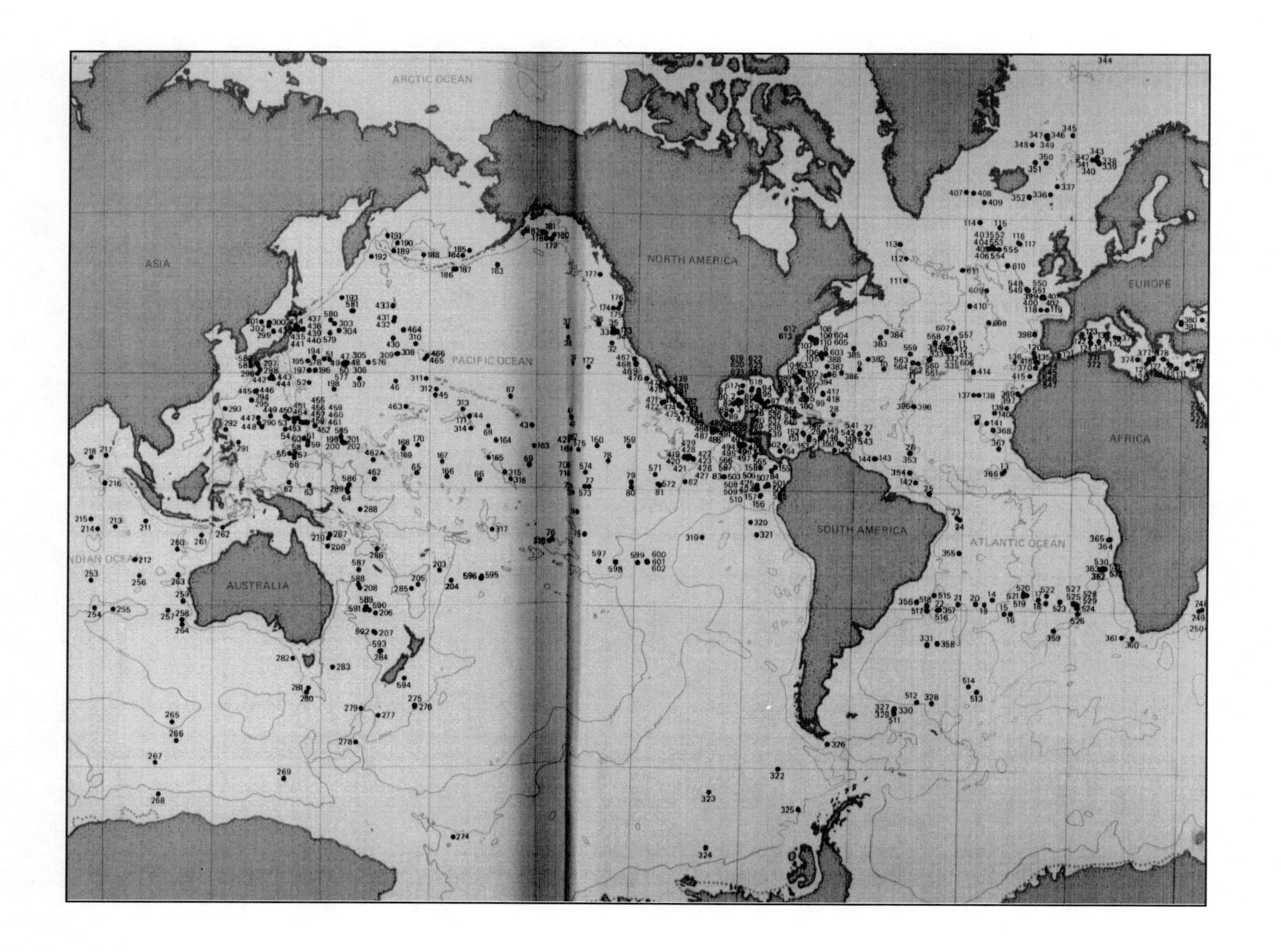
ARCTIC OCEAN
ASIA
NORTH AMERICA
EUROPE
PACIFIC OCEAN
AFRICA
SOUTH AMERICA
ATLANTIC OCEAN
AUSTRALIA
NDIAN OCEA

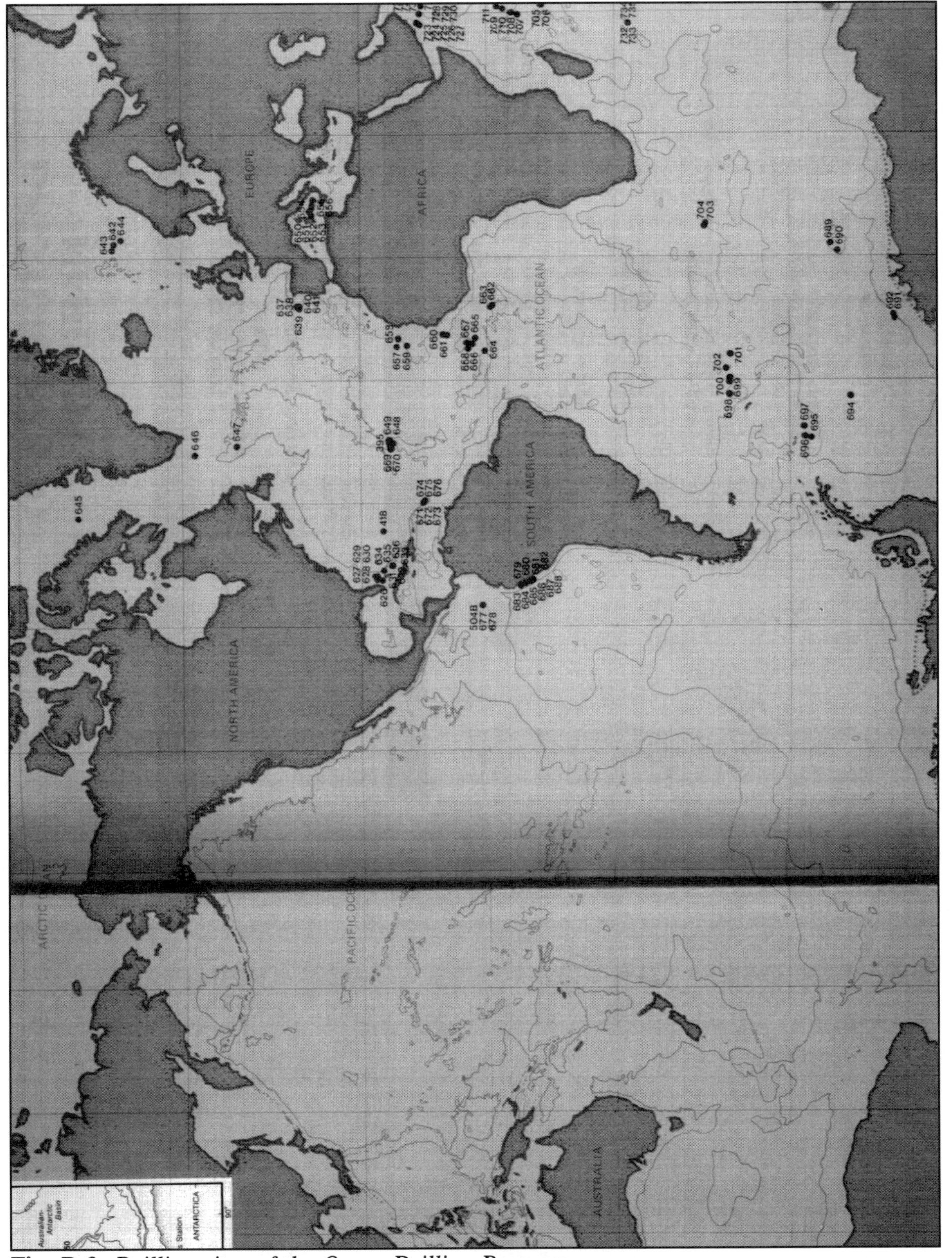

Fig. B.3 *Drilling sites of the Ocean Drilling Program.*

Equipment

Fig. B.4 *A section of drill pipe being readied for use on the JOIDES Resolution.*

The designers of the drilling ships *Glomar Challenger* and ***JOIDES Resolution*** made sure that the 400-foot long ships were heavy for their size and innately stable. Because some heaving in the seas was inevitable, they provided telescoping sections of drill pipe to hold the bit steady as it ground into the sea floor. One of the most innovative designs was a computer-controlled positioning system. By compensating for wind and currents with main and thruster propellers, the system holds the ship nearly motionless over the drill hole while drilling in water depths of up to 27,000 feet, even in storms so severe that the main deck is sometimes awash. The ***JOIDES Resolution*** has the ability to suspend up to 30,000 feet of drill pipe (See Fig. B.4).

Fig. B.5 *The advanced piston corer. With this new tool, undisturbed 2.5 inch diameter core samples can be recovered in soft sediments as deep as 250 meters beneath the sea floor.*

The engineering staff has developed tools and techniques to recover core samples from geologic formations around the world. The following suite of tools is the workhorse system of the program. Most **ODP** holes are drilled with various combinations of these tools. The tools are designed to penetrate progressively into more difficult layers of ocean crust.

Advanced Piston Corer: This tool recovers undisturbed core in soft ooze and sediments at the seafloor. The system references the core to magnetic north and can measure temperatures in the hole. The system can usually recover core as deep as 250 meters below the seafloor. Figure B.5 shows the Advanced Piston Corer and Figure B.6 shows a comparison between cores recovered with and without the Advanced Piston Corer.

Extended Core Barrel: This tool continues coring in firm sediments after piston coring is no longer possible.

Hard-Rock Guide Base: This tool allows the bit to drill or "spud" into fractured basalts that are devoid of a sediment cover. With metal weights adding ballast, the submerged weight of this base is over 100,000 pounds and holds the drill in place until it penetrates the rock. A "funnel" at the top of the base guides the drill to the selected site.

Diamond Coring System: This tool is designed to core and recover both fractured hard-rock formations as well as alternating hard and soft formations such as chert and chalk. A slim string of high-strength tubing is deployed through the drillstring and rotated ahead of the large bit. A specially designed drilling platform rides in the derrick.

The Cork: This borehole seal allows instruments to be placed down the hole to monitor various measurements including temperature and pressure. Remotely operated vehicles such as Alvin can latch onto the grating and extract measurements. Drilling can be initiated again later at such a site, if desired. Figure B.7 shows a Cork installed on a site at the Juan de Fuca Ridge.

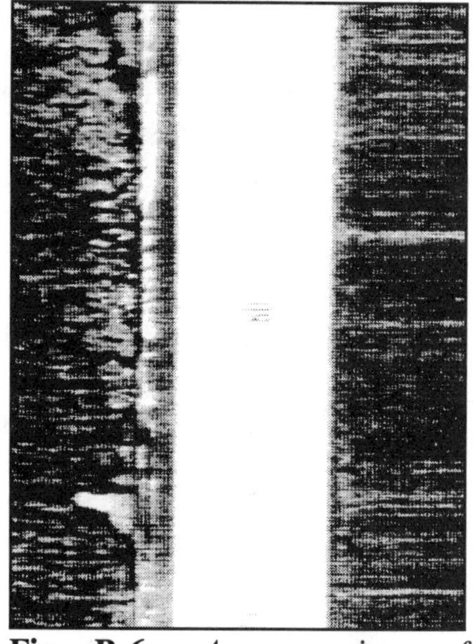

Fig. B.6 *A comparison of sediment cores recovered by the standard rotary coring method (left) with the advanced piston corer method (right).*

Many of the drilling systems use standard mining technology applied to a floating drill ship. For example, the Diamond Coring System, the Vibra-Percussive Corer, the Sonic Core Monitor, and the drill tower and crew procedures were all adapted from standard mining technology. Figure B.8 shows a drill crew making up a stand of drill collars before tripping the pipe.

Fig. B.7 *The "cork" used to seal a borehole and install instruments down the hole for monitoring temperature and pressure.*

On-board Research Facilities

The *JOIDES Resolution* provides twelve laboratories on seven levels of the ship for studies in sedimentology, paleontology, petrology, geochemistry, geophysics, paleomagnetics and physical properties. Photographs are taken of all cores to record their physical properties. Specialized instruments include X-ray diffraction and X-ray fluorescence machines to analyze the mineral structure and chemical composition of sediment and rock samples. Figure B.9 shows a cryogenic magnetometer which determines a cores's magnetization, used to record the history of magnetic reversals in the earth's magnetic field. Electronics, photographic, library, and computer facilities provide technical support. A fully-computerized system of recording and checking the data is used throughout the acquisition and processing of cores.

Fig. B.8 *A drill crew making up a stand of drill collars before "tripping" the pipe.*

The cores are cut in 1.5 meter sections, split in half lengthwise, marked according to their location and depth, and photographed. Half of the core is immediately placed in storage and the other half (the working half) run through a series of on-board analyses. The set of on-board analyses includes the following:

Sediment Data:
- Carbon/Carbonate
- Density-Porosity
- Grain Size
- GRAPE (Gamma Ray Attenuation Porosity Evaluator)
- Paleontology
- Pore Water
- SCREEN
- Smearslide
- Sonic Velocity
- Vane Shear
- Visual Text
- X-Ray Mineralogy

Igneous/Metamorphic Rock Data
- Major Elements
- Minor Elements
- Thin Sections
- Visual Core Description

Paleomagnetics
- AFD Sediment
- Discrete Sediment
- Long-core Sediment
- Hard Rock

Reference
- Age Profile
- Core Depths

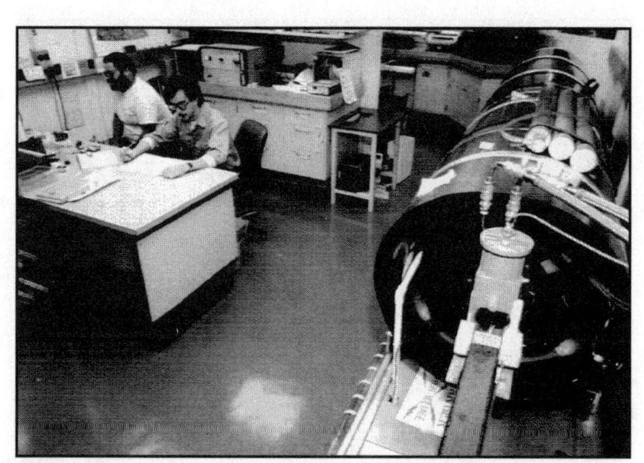

Fig. B.9 *The cryogenic magnetometer on the JOIDES Resolution used to determine a core's magnetization.*

Samples are taken from the working half of the core for staff scientists and other investigators who wish to do processing in their own labs. The total volume of samples removed during a cruise is not allowed to exceed one half of the working half of the core. No interval is allowed to be depleted. The archive half, which is never sampled, can be studied by non-destructive techniques only.

ODP Accomplishments

By the end of the Deep Sea Drilling Project, the *Glomar Challenger* had completed 96 internationally-staffed expeditions. The ship had drilled primarily in the Atlantic, Pacific, and Indian Oceans. Table B.1 lists the legs, locations, repositories, and sites of **DSDP**. The sites can be located on the map of Figure B.2.

By the end of 1990, ***JOIDES** Resolution* had completed 34 internationally staffed expeditions. The ship had drilled extensively in the Atlantic, Pacific, and Indian oceans. ***JOIDES** Resolution* has explored the Mediterranean, Norwegian, Labrador, Weddell, Sulu, Celebes, Philippine, and Japan seas. It has successfully tested its high-latitude capabilities north of the Arctic and south of the Antarctic circles. Table B.2 lists the legs, locations, repositories, and sites of the **ODP**. The sites can be located on the map of Figure B.3.

Approximately 800 scientists sailing on 34 cruises have addressed scientific objectives relating to the history of ancient oceanic environments, and the origin of ocean crust, continental margins, and sediment layers. **ODP** has also documented changes in earth's oceans, atmosphere, polar regions, biosphere, and magnetic fields. During these cruises, **ODP** has obtained nearly 70 miles of sediment and rock for geologic research.

Organization of the Ocean Drilling Program

The U.S. National Science Foundation funds the program with significant contributions from 19 member countries. The Joint Oceanographic Institutions Inc. manages the program. The Joint Oceanographic Institutions for Deep Earth Sampling (**JOIDES**) provides overall scientific advice. Members of **JOIDES**, a worldwide network of universities, oceanographic institutions and government agencies, are dedicated to the advancement of marine geology and oceanography through scientific ocean drilling.

Members of the Ocean Drilling Program are:

University of California at San Diego
Columbia University
University of Hawaii
University of Miami
Oregon State University
University of Rhode Island
Texas A&M University

University of Texas at Austin
University of Washington
Woods Hole Oceanographic Institution
Canada/Australia Consortium for the **ODP**
European Science Foundation Consortium for the **ODP**
Belgium, Denmark, Finland, Iceland, Italy,
Greece, the Netherlands, Norway, Spain, Sweden, Switzerland, and Turkey
Federal Republic of Germany
France
Japan
United Kingdom
U.S.S.R.

Publications of the Ocean Drilling Program

The primary publications of the **ODP** are the Initial Reports and the Scientific Results. These publications are written and edited by shipboard scientists and distributed by the Ocean Drilling Program at Texas A&M University under contract with the National Science Foundation.

The Initial Reports describe the collection procedures, initial analyses, and preliminary results of a particular leg. Initial Reports are distributed promptly to facilitate rapid information exchange, generally within a year or so after a cruise has been completed.

Scientific Results describe in-depth analyses and interpretation not possible on-board. For example, most of the $\delta^{18}O$ data are collected from samples after a cruise and reported in the Scientific Results. It may take several years for the Scientific Results from a particular cruise to be distributed.

The Reports carry the same volume number as the number of the leg (a leg or cruise will normally drill at several sites). The status of forthcoming reports is listed in the frontispiece of each volume and in a newsletter distributed by **JOIDES**. Copies of the reports were $45.00 each in January of 1993 and may be obtained from:

Publications Distribution Center
Ocean Drilling Program
1000 Discovery Drive
College Station, TX 77845-9547

Table B.1 Drilling Locations for the Deep Sea Drilling Project

Leg	Location	Rep.	Sites
1	N. Atlantic	E	1-7
2	N. Atlantic	E	8-12
3	S. Atlantic	E	13-22
4	Cntrl. Atlan.	E	23-31
5	N. Pacific	W	32-43
6	N. Pacific	W	44-60
7	Cntrl. Pac.	W	61-67
8	Cntrl. Pac.	W	68-75
9	Cntrl. Pac.	W	76-84
10	Gulf of Mex.	E	85-97
11	N. Atlantic	E	98-108
Re-entry Trials		E	109-110
12	N. Atlantic	E	111-119
13	Mediterranean	E	120-134
14	Cntrl. Atlan.	E	135-144
15	Caribbean	E	146-154
16	Cntrl. Pac.	W	155-163
17	Cntrl. Pac.	W	164-171
18	N. Pacific	W	172-182
19	N. Pacific	W	183-193
20	N. Pacific	W	194-202
21	Coral Sea	W	203-210
22	Indian Ocean	W	211-218
23	Arabian Sea	W	219-230
24	Indian Ocean	W	231-238
25	Indian Ocean	W	239-249
26	Indian Ocean	W	250-258
27	Indian Ocean	W	259-263
28	Antarctic	E	264-274
29	Tasman Sea	E	275-284
30	S. Pacific	W	285-289
31	N. Pacific	W	290-302
32	N. Pacific	W	303-313
33	N. Pacific	W	314-318
34	S. Pacific	W	319-321
35	S. Pacific	E	322-325
36	S. Pacific	E	326-331
37	N. Atlantic	E	332-335
38	Arctic	E	336-352
39	Atlantic	E	353-359
40	S. Atlantic	E	360-365
41	S. Atlantic	E	366-370
42A	Mediterranean	E	371-378
42B	Black Sea	E	379-381
43	N. Atlantic	E	382-387
44	N. Atlantic	E	388-392
44A	N. Atlantic	E	393-394
45	N. Atlantic	E	395-396
46	N. Atlantic	E	396A-396B
47	N. Atlantic	E	397-398
48	N. Atlantic	E	399-406
49	N. Atlantic	E	407-414
50	N. Atlantic	E	415-416
51A	N. Atlantic	E	417-417A
51B	N. Atlantic	E	417B-417D
52	N. Atlantic	E	417D-418A
53	N. Atlantic	E	418A-418B
54	Cntrl. Pac.	W	419-429
55	N. Pacific	W	430-433
56	N. Pacific	W	434-437
57	N. Pacific	W	438-441
58	N. Pacific	W	442-446
59	N. Pacific	W	447-451
60	N. Pacific	W	452-461
61	Cntrl. Pac.	W	462
62	N. Pacific	W	463-466
63	N. Pacific	W	467-473
64	Gulf of Calif	W	474-481
65	Gulf of Calif	W	482-485
66	Cntrl. Pac.	W	486-493
67	Cntrl. Pac.	W	494-500
68	Cntrl. Pac.	W	502-503
69	Cntrl. Pac.	W	501,504-505
70	Cntrl. Pac.	W	504B,506-510
71	S. Atlantic	E	511-514
72	S. Atlantic	E	515-518
73	S. Atlantic	E	519-524
74	S. Atlantic	E	525-529
75	S. Atlantic	E	530-532
76	N. Atlantic	E	533-534
77	Caribbean	E	535-540
78	N. Atlantic	E	541-543,395A&B
79	N. Atlantic	E	544-547
80	N. Atlantic	E	548-551
81	N. Atlantic	E	552-555
82	N. Atlantic	E	556-564
83	Cntrl. Pac.	W	504B
84	Cntrl. Pac.	W	565-570
85	Cntrl. Pac.	W	571-575
86	N. Pacific	W	576-581
87	N. Pacific	W	582-584
88	N. Pacific	W	581
89	Cntrl Pac.	W	462A,585-586
90	S. Pacific	W	587-594
91	Cntrl. Pac.	W	595-596
92	Cntrl. Pac.	W	504B,597-602
93	N. Atlantic	E	603-605
94	N. Atlantic	E	606-611
95	N. Atlantic	E	603,612-613
96	Gulf of Mex	E	614-624

Core Storage
E - East Coast Repository
W - West Coast Repository

Table B.2 Drilling Locations for the Ocean Drilling Project

Leg	Location	Rep.	Sites
100	Gulf of Mex.	E	625
101	Bahamas	E	626-636
102	Bermuda Rise	E	418
103	Galicia Margin	E	637-641
104	Norwegian Sea	E	642-644
105	Baffin Bay-Labrador Sea	E	645-647
106	Mid-Atlantic Ridge	E	648-649
107	Tyrrhenian Sea	E	650-656
108	Eastern Tropical Atlantic	E	657-668
109	Mid-Atlantic Ridge	E	395,648,669-670
110	Northern Barbados Ridge	E	671-676
111	Costa Rica Rift	G	504,677-678
112	Peru Continental Margin	G	679-688
113	Weddell Sea, Antarctica	E	689-697
114	Subantarctic S. Atlantic	E	698-704
115	Mascarene Plateau	G	705-716
116	Distal Bengal Fan	G	717-719
117	Oman Margin	G	720-731
118	Southwest Indian Ridge	G	732-735
119	Kerguelen Plateau-Prydz Bay	E	736-746
120	Central Kerguelen Plateau	E	747-751

APPENDIX C

GENERATION OF STABLE OXYGEN ISOTOPE DATA

An incredible amount of data has been collected on the various characteristics of sea-floor sediments world-wide. Some of the variables measured and the global distribution of sampling were shown in appendix C. Thousands of scientists are involved in the analysis and interpretation of these data. This report will treat only one primary variable of interest — the concentration of oxygen isotopes.

The primary reason for selecting this variable is the relation it has to temperature. Variations in the relative concentrations of oxygen isotopes are associated with the temperature at which micro-organisms form their skeletal and shell structures. By a proper analysis of oxygen isotope ratios in sea-floor sediments a chronology of ocean temperature and global climate can be inferred. These isotope ratios have been measured in ice cores taken from Greenland and Antarctica. There appears to be an inverse relation between the relative concentration of oxygen isotopes in ice sheets and in sea-floor sediments, allowing a model to be developed for interactions among the hydrosphere, biosphere, cryosphere, and atmosphere.

Laboratory Procedures

The acquisition of stable isotope data begins with sampling from a selected core. Samples are taken at intervals along the core. The quantity of a carbonate sample needed for analysis is usually 0.1-1.0 mg. If the analysis is to be done on a bulk sample, a large enough sample need only be taken to allow for removal of the non-carbonate material. If the analysis is to be done only on a given species of micro-organism, the sample must be large enough to contain .1 - 1.0 mg of the particular species. Carbonate samples may be in the form of whole rock, forams, calcareous nannofossils, or carbonate grains of detrital origin, depending on the depositional basin, and the time of deposition.

The carbonate material is first roasted at 380°C in a vacuum to remove any organic contaminants. Figure C.1 shows examples of various vacuum ovens used to roast carbonates. Samples contaminated with drilling mud, or containing large quantities of hydrocarbons, as in an oil shale, may require a special pretreatment. After removal of organic material, the carbonate fraction is treated with purified phosphoric acid (H_3PO_4) in an evacuated reaction vessel held at a constant temperature of 50, 60, or 70°C, depending on the reaction time required to achieve total conversion of all $CaCO_3$ to $Ca_3(PO_4)_2$. As indicated in the chemical reaction Equation C.1, the oxygen in the $CaCO_3$ is transferred to H_2O and CO_2.

$$2\ H_3PO_4 + 3\ CaCO_3 \rightleftharpoons Ca_3(PO_4)_2 + 3\ H_2O\uparrow + 3\ CO_2\uparrow \quad \text{(C.1)}$$

The resultant carbon dioxide gas formed in this process is separated from the water vapor by fractional freezing using liquid nitrogen and transferred to an isotope ratio mass spectrometer. Figure C.2 shows an example of a mass spectrometer. In the mass spectrometer, the CO_2 molecules are ionized and separated into ion beams of three different masses: 44 ($^{12}C^{16}O^{16}O$), 45 ($^{13}C^{16}O^{16}O$), and 46 ($^{12}C^{16}O^{18}O$). The ratio of the intensities of the 46 and 44 ion beams in the mass spectrometer yields the $^{18}O/^{16}O$ ratio of the CO_2 gas. The ratio of the 45 and 44 ion beams yields the $^{13}C/^{12}C$ ratio, an important variable for estimating changes in carbon distributions between the ocean, atmosphere, and biosphere. The $^{18}O/^{16}O$ and $^{13}C/^{12}C$ ratios of the sample CO_2 are specified in terms of the corresponding ratios in a laboratory reference CO_2 gas as measured under identical operating conditions. By convention, the ratio is then expressed as a delta (δ) value in parts per thousand (per mil, ‰) enrichment or depletion in the minor isotope, ^{18}O or ^{13}C, respectively:

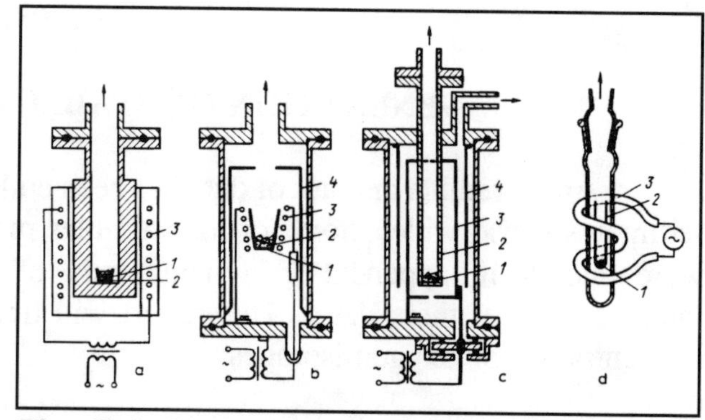

Fig. C.1 *Devices used to heat solid samples in a vacuum. Heating can be done internally or externally to the vacuum. The arrows lead to an analytic instrument.*

$$\delta^{18}O = \frac{[^{18}O/^{16}O]_{sample} - [^{18}O/^{16}O]_{reference}}{[^{18}O/^{16}O]_{reference}} X\ 1000 \qquad \textbf{(C.2)}$$

$$\delta^{13}C = \frac{[^{13}C/^{12}C]_{sample} - [^{13}C/^{12}C]_{reference}}{[^{13}C/^{12}C]_{reference}} X\ 1000 \qquad \textbf{(C.3)}$$

Positive delta values specify more ^{18}O or ^{13}C than in the reference CO_2. For example, a typical mass ratio of $^{18}O/^{16}O$ for a reference may give a value of $1x10^{-6}$. If a given sample of interest gives, say, $1.001x10^{-6}$, then the difference is $1x10^{-9}$ and the relative difference is $1x10^{-3}$. When multiplied by 1000, as shown in Eqs. C.2 and C.3, the $\delta^{18}O$ would be +1‰ for this sample. If the sample has a smaller mass ratio than that of the reference standard, the delta values become negative.

The isotope ratio mass spectrometers now available can determine $\delta^{18}O$ and $\delta^{13}C$ values with a precision of ±.02‰, which is less than 2% of the total range of isotope variation observed in biogenic carbonate from Tertiary sediments. To guarantee that delta values can be compared between laboratories, every major isotope laboratory corrects the measured delta values for spectrometer differences (Craig, 1957; Deines, 1970), and calibrates its reference CO_2 gas to the universal PDB (Peedee Belemnite) standard. All values for forams are reported relative to PDB, which is powdered belemnite from the Peedee Formation of South Carolina (Urey et al., 1951; Epstein et al., 1953). To ensure the integrity of the isotopic data, a laboratory reference carbonate is analyzed daily prior to the analysis of each set of unknown carbonates.

Fig. C.2 *A modern solid-source mass spectrometer. At the right side of the photograph is a large electromagnet that deflects the beam separating the ions by mass.*

Isotopic Temperature

The technique most widely and successfully applied to the study of Cenozoic climates in the past several years has been the oxygen isotope paleotemperature method of Urey et al. (1951). This technique is based on the fact that there is a difference between the isotope ratio in calcium carbonate and in water associated with its formation. This difference is a function of depositional temperature. Therefore, by measuring $^{18}O/^{16}O$ in fossil carbonate shells, estimating $^{18}O/^{16}O$ of the water in which they grew, and knowing the manner in which the isotopic fractionation between calcium carbonate and water varies with temperature, it is possible to estimate the temperature at which the carbonate was deposited. Temperature estimates made in this way are referred to as isotopic temperatures. Measurements of $^{18}O/^{16}O$ made with a precision of $\pm$.02‰ correspond to a precision in the temperature estimate of approximately 0.1°C. The precautions necessary for valid applications of the technique and interpretation of results have been discussed by several authors, including Urey et al. (1951), Bowen (1966), and Savin, Douglas and Stehli (1975).

A record of ancient seawater temperature may be obtained from certain fossil $^{18}O/^{16}O$ ratios. Many organisms secrete calcium carbonate either in isotopic equilibrium (equal $^{18}O/^{16}O$ ratio) with seawater or very close to equilibrium. Fossils of such organisms identify the $^{18}O/^{16}O$ ratio of seawater at the time these organisms lived. Mollusks and forams have been most frequently used in Cenozoic paleotemperature studies. The attainment of isotopic equilibrium has been carefully determined for planktic forams (species of forams which live in approximately the upper 100 meters of the ocean, called the photic zone). There is some indication that planktic forams may exhibit small departures from equilibrium, especially at tropical temperatures (van Donk, 1970; Shackleton, Wiseman & Buckley, 1973). While further investigation of this is desirable, it is most likely that these small departures from isotopic equilibrium in planktic forams will be of greater importance in studies of foram biology and ecology than in paleotemperature studies.

Benthic forams (See Figure C.3 for an example of a benthic foram which lives at or near the ocean floor) in deep-sea sediments frequently exhibit departures from isotopic equilibrium (Duplessy, LaLou & Vinot, 1970; Shackleton, 1974; Woodruff, unpublished data). Shackleton (1974) reported that a type of foram called Uvigerina deposits its shell at or near isotopic equilibrium in the temperature range 0.8°C to 7°C. The magnitude of disequilibrium in oxygen isotope fractionation for different types of benthic forams remains relatively constant throughout single cores, permitting empirical intercalibration of the isotopic temperatures obtained from different types. There is evidence that this intercalibration is not always simple.

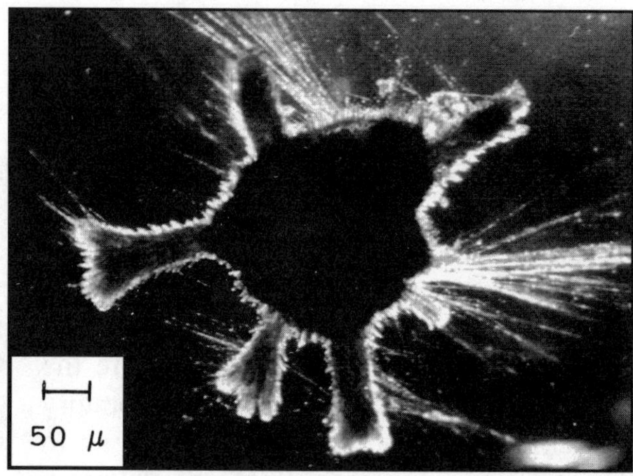

Fig. C.3 *A living benthic foram from the West Pacific. The protrusians, called pseudopodia, emerge from spines in the shell.*

The isotopic composition of biogenic calcium carbonate must remain unaltered after deposition if it is to have paleoclimatic significance. Changes can occur primarily by chemical processes in sediments between deposition and conversion into rock (lithification). This low-temperature, relatively low-pressure alteration process is called diagenesis. See Figure C.4 for an example of the type of data used to detect recrystallization. Evidence for recrystallization or chemical alteration may be taken as sufficient grounds for the elimination of a specimen from an isotopic paleo-temperature study (Urey et al., 1951). Criteria for recognizing an acceptable state of preservation have been reviewed by Stevens and Clayton (1971) and Spaeth, Hoefs and Vetter (1971).

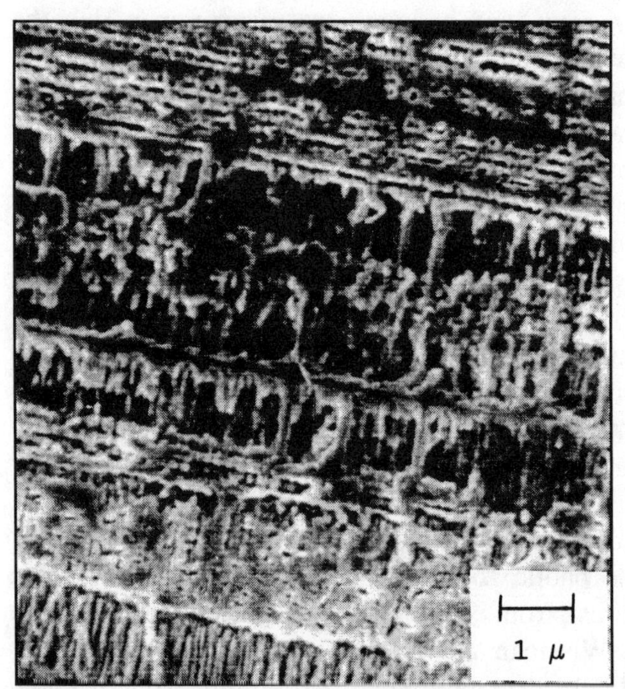

Fig. C.4 *A scanning electron micrograph of an etched section of the shell of a benthic foraminifer. Analysis of the material in the layers is used to detect recrystalization.*

In addition to crystallization, other diagenetic processes can cause important modifications to the isotopic temperature record of materials from deep-sea sediments. Species-related requirements restrict the range of water depths occupied by individuals of a single species of planktic forams living at a single location. Within that range of depths, however, individuals may form with a range of isotopic compositions corresponding to the *in situ* temperature variation.

Shallow-dwelling individuals of a species are probably more subject to solution, a process ubiquitous in deep marine basins. Preferential loss of the "warmer" fraction of a foram assemblage through selective solution biases isotopic temperatures of the remaining forams to lower values (Savin and Douglas, 1973). This effect may be as much as 2 or 3°C in recent tropical sediments. In the tropics the effect is greater for shallower-dwelling species of forams and is more pronounced for samples in sediments from deeper waters. The importance of this solution effect is probably less for Tertiary and Cretaceous sediments than for recent ones, since during those times the thermocline (an oceanic water layer in which water temperature decreases rapidly with increasing depth) in tropical regimes was less pronounced than at present (Savin, Douglas and Stehli, 1975).

There are also ecological factors that must be considered in the interpretation of isotopic paleotemperature data. Even assuming that temperatures of carbonate deposition can be accurately estimated from isotopic data, these temperatures do not have quantitative climatic significance until interpreted further. In regions with seasonal temperature variations, carbonate precipitation may occur predominantly at one time of the year, as pointed out for mollusks by Epstein and Lowenstam (1953). Different types of forams occupy water of different average depths and hence different temperatures. Additionally, individuals of many types migrate into waters of different depths or temperatures either seasonally or as a function of growth stage. Temperatures of carbonate deposition do not, therefore, in general indicate temperatures at the ocean surface. However, surface temperatures can frequently be estimated from isotopic temperatures when the growth habits of the organisms are understood.

The depth stratification of planktic forams and its expression in the $^{18}O/^{16}O$ ratio of their calcite was discussed initially by Emiliani (1954) and subsequently by Lidz, Kehm,and Miller (1968); Shackleton (1968); Oba (1969); Emiliani (1971); Hecht and Savin (1972); Duplessy (1972); Savin and Douglas (1973); Savin and Stehli (1974); and others. Depth habitats of extinct Tertiary and Upper Cretaceous planktic forams have been reconstructed using isotopic evidence (Douglas and Savin, 1977). Extinct types of forams that are morphologically similar to present-day planktic species are believed to exhibit depth stratification similar to their modern counterparts. However, even modern planktic forams exhibit significant changes which affect the $^{18}O/^{16}O$ ratio. For example, forams change depths during their life cycle. During gametogenesis, they often sink and gain a fair amount of $CaCO_3$.

This information on habitats of forams, when combined with data on the differences between surface temperatures and isotopic temperatures of forams from low-latitude recent ocean sediments has been used to infer Tertiary surface temperatures from the isotopic temperatures of Tertiary forams (Savin and Douglas, 1973; Savin, Douglas and Stehli, 1975). Such inferences, however, are based on speculative assumptions concerning the past thermal structure of the upper portion of the water column. The relationship between isotopic temperatures of forams and surface temperatures requires further exploration.

To relate temperatures of carbonate secretion to surface temperatures and especially to the surface temperature fluctuations between glacial and postglacial times, it is necessary to understand how the depth habitats of forams change as climatic conditions change. Every species of living forams appears to preferentially inhabit water of a particular density (Emiliani, 1954). If the density preferences of the species have remained constant from glacial to postglacial periods, dilution of the oceans by icecap runoff and the resultant lowering of the density of seawater would result in the downward migration of forams. Conversely, icecap growth would result, other things being equal, in upward migration of forams. The net effect for forams that migrate in this fashion would be to minimize glacial-postglacial differences in isotopic temperatures (Shackleton, 1968). Savin and Douglas (1973) suggested that osmotic interaction with the surrounding water is an important factor in determining the depth habitats of forams. It has also been found that other organisms living in close association limit many planktic and some benthic forams to the photic zone. If these and other factors such as the availability of nutrients are important in determining their depth habitats, the vertical migration of forams between glacial and postglacial times may be more complex than suggested by simple density models (Savin and Stehli, 1974).

Finally, in determining temperatures of shell secretion from isotopic data, $\delta^{18}O$ of the water in which the shell grew must be estimated. Compared with fresh waters, the oxygen isotopic compositions of ocean waters of normal salinity fall within a fairly narrow range (about 2.5 ppmil). Within this range, $\delta^{18}O$ is affected primarily by processes of evaporation and precipitation, and in certain regions, influx of meltwater (Epstein and Mayeda, 1953). These factors and the manner in which they affect the isotopic composition of the modern oceans have been discussed in Craig and Gordon (1965).

In the interpretation of isotopic paleotemperature data, variation of the isotopic composition of seawater from two processes must be considered: 1) Large-scale polar icecap formation results in the preferential removal of ^{16}O from the oceans into the icecaps (^{16}O is more readily evaporated than ^{18}O), and 2) the corresponding enrichment of the oceans in ^{18}O. The periodic advance and retreat of polar icecaps during the Pleistocene is associated with corresponding periodic fluctuations in $\delta^{18}O$ of the oceans. The magnitude of these fluctuations remains one of the most controversial topics in the interpretation of Pleistocene isotopic paleotemperature data. As our picture of past oceanographic and climatic conditions becomes more complete it should become possible to include data which will account for local variations in oceanic $\delta^{18}O$.

The isotopic compositions of many organically precipitated carbonates contain climatic information, obscured only in part by the factors discussed above. The more accurately the effects other than temperature can be determined, the more realistically the isotopic data can be interpreted. Given all of the uncertainties involved and the catastrophic changes in ocean temperature and habitats that have probably occurred, as the Flood model suggests, it is unlikely that ancient surface temperatures can be estimated from isotopic data with an accuracy better than 3 or 4°C.

APPENDIX D

THE CONVENTIONAL GEOLOGIC TIME SCALE

Geologists have studied the crust of the earth in great detail and have classified the rock layers and sediments into major and minor groupings. Since most geologic processes create newer layers on top of older layers, the normal sequence of events, in either an evolutionary or creationist framework, occurs from bottom to top in the geologic column. There are obvious exceptions to this sequence when layers are reworked, *e.g.* newer layers are formed under older layers, or older layers are forced over newer layers. In general, the sequence of events is not in dispute between most creationists and evolutionists.

There are major disagreements over the length of time required to form a given layer or the period of time between layers, however. The conventional evolutionary time scale extends over 4.5 billion years, while the Biblical creationist time scale is on the order of 10,000 years. This disagreement implies major differences in processes and process rates. This issue is at the crux of the work behind this report. The evolutionary model suggests that the rock layers and sediments were generally laid down slowly by low-energy, uniformitarian processes. The creationist model suggests they were primarily laid down rapidly by high-energy, catastrophic processes.

Before a detailed presentation of the different models of sediment formation are presented, it is useful to describe the subdivisions of geological strata and terminology that will be used in the discussion. Fig. D.1 shows the Geologic Time Scale published by the Geological Society of America. This time scale generally matches the sequence of events in the time scales that have been developed over the last century or so by the conventional geological community. The assigned times and periods have increased as the presumed age of the earth has been extended. This scale is useful for our discussion because it contains the standard terminology used to refer to layers and periods of time in earth history. The times assigned to the layers are used extensively in the evolutionary model. The same terminology for the layers and events will be used for the creationist model in this report, but major adjustments will be made to the time frames.

Earth history is conventionally broken into four major eras, from the oldest to the most recent: (1) Precambrian, (2) Paleozoic, (3) Mesozoic, and (4) Cenozoic. Precambrian layers occur below the Cambrian Period found in the Paleozoic era in which widely varied and extensive life forms are evident from their fossil remains. According to the conventional evolutionary time scale, the Precambrian era lasted over 4 billion years. The Paleozoic Era (early life), the Mesozoic Era (middle life), and the Cenozoic Era (recent life) form three subdivisions of what is sometime referred to as the Phanerozoic (visible life).

Each era is broken into finer periods and epochs. The periods of the Paleozoic are the Cambrian, Ordovician, Silurian, Devonian, Carboniferous (with the Mississippian and Pennsylvanian subdivisions), and Permian. The periods of the Mesozoic are the Triassic,

DNAG

DECADE OF NORTH AMERICAN GEOLOGY
GEOLOGIC TIME SCALE

GEOLOGICAL SOCIETY OF AMERICA

CENOZOIC

PERIOD		EPOCH		AGE	PICKS (Ma)
QUATERNARY		HOLOCENE			0.01
		PLEISTOCENE		CALABRIAN	1.6
TERTIARY	NEOGENE	PLIOCENE	L	PIACENZIAN	3.4
			E	ZANCLEAN	5.3
		MIOCENE	L	MESSINIAN	6.5
				TORTONIAN	11.2
			M	SERRAVALLIAN	15.1
				LANGHIAN	16.6
			E	BURDIGALIAN	21.8
				AQUITANIAN	23.7
	PALEOGENE	OLIGOCENE	L	CHATTIAN	30.0
			E	RUPELIAN	36.6
		EOCENE	L	PRIABONIAN	40.0
			M	BARTONIAN	43.6
				LUTETIAN	52.0
			E	YPRESIAN	57.8
		PALEOCENE	L	SELANDIAN: THANETIAN	60.6
				UNNAMED	63.6
			E	DANIAN	66.4

MESOZOIC

PERIOD	EPOCH	AGE	PICKS (Ma)	UNCERT. (m.y.)
CRETACEOUS	LATE		66.4	
		MAASTRICHTIAN	74.5	4
		CAMPANIAN	84.0	4.5
		SANTONIAN	87.5	
		CONIACIAN	88.5	2.5
		TURONIAN	91	
		CENOMANIAN	97.5	2.5
	EARLY	ALBIAN	113	4
		APTIAN	119	9
	NEOCOMIAN	BARREMIAN	124	9
		HAUTERIVIAN	131	8
		VALANGINIAN	138	5
		BERRIASIAN	144	5
JURASSIC	LATE	TITHONIAN	152	12
		KIMMERIDGIAN	156	6
		OXFORDIAN	163	15
	MIDDLE	CALLOVIAN	169	15
		BATHONIAN	176	34
		BAJOCIAN	183	34
		AALENIAN	187	34
	EARLY	TOARCIAN	193	28
		PLIENSBACHIAN	198	32
		SINEMURIAN	204	18
		HETTANGIAN	208	18
TRIASSIC	LATE	NORIAN	225	8
		CARNIAN	230	22
	MIDDLE	LADINIAN	235	10
		ANISIAN	240	22
	EARLY	SCYTHIAN	245	20

RAPID POLARITY CHANGES

PALEOZOIC

PERIOD		EPOCH	AGE	PICKS (Ma)	UNCERT. (m.y.)
PERMIAN		LATE	TATARIAN	245	20
			KAZANIAN	253	20
			UFIMIAN	258	24
			KUNGURIAN	263	22
		EARLY	ARTINSKIAN	268	12
			SAKMARIAN		
			ASSELIAN	286	12
CARBONIFEROUS	PENNSYLVANIAN	LATE	GZELIAN		
			KASIMOVIAN	296	10
			MOSCOVIAN		
			BASHKIRIAN	315	20
	MISSISSIPPIAN			320	
		EARLY	SERPUKHOVIAN	333	22
			VISEAN		
			TOURNAISIAN	352	8
DEVONIAN		LATE	FAMENNIAN	360	10
			FRASNIAN	367	12
		MIDDLE	GIVETIAN	374	18
			EIFELIAN	380	18
		EARLY	EMSIAN	387	28
			SIEGENIAN	394	22
			GEDINNIAN	401	18
SILURIAN		LATE	PRIDOLIAN	408	12
			LUDLOVIAN	414	12
		EARLY	WENLOCKIAN	421	12
			LLANDOVERIAN	428	8
ORDOVICIAN		LATE	ASHGILLIAN	438	12
			CARADOCIAN	448	12
		MIDDLE	LLANDEILAN	458	16
			LLANVIRNIAN	468	16
		EARLY	ARENIGIAN	478	16
			TREMADOCIAN	488	20
CAMBRIAN		LATE	TREMPEALEAUAN	505	32
			FRANCONIAN		
			DRESBACHIAN	523	38
		MIDDLE		540	28
		EARLY		570	

PRECAMBRIAN

EON	ERA	BDY. AGES (Ma)
PROTEROZOIC	LATE	570
	MIDDLE	900
	EARLY	1600
ARCHEAN	LATE	2500
	MIDDLE	3000
	EARLY	3400
		3800?

Compiled 1983

Published by The Geological Society of America, Inc. 3300 Penrose Place, P.O. Box 9140 Boulder, Colorado 80301

MAP AND CHART SERIES MC-50

Fig. D.1 *The current geologic time scale published by the Geological Society of America.*

Jurassic, and Cretaceous. The periods of the Cenozoic are the Tertiary and the Quaternary, although sometimes shown as the Paleogene, Neogene, and Quaternary. These periods are often broken into epochs; normally early, middle, and late; and even shorter ages. However, since I will be concentrating on the upper portion of the record, I will list only the epochs for the Cenozoic. They are the Paleocene, Eocene, Oligocene, Miocene, Pliocene, Pleistocene, and Holocene. The Pleistocene Epoch extends back 1.6 million years before present and contains evidence for numerous "Ice Ages" according to the conventional time scale. The Holocene Age covers the most recent 10,000 years since the last "Ice Age."

Fig. D.2 *An example of the Cretaceous/Tertiary boundary. The band of clay exposed near Gubbio, Italy contains a concentration of iridium 30 times higher than in sediments above and below.*

A significant boundary is evident between the Cretaceous Period in the Mesozoic Era and the Tertiary Period in the Cenozoic Era. Catastrophic events which occurred at this time in earth history are evident even to the most devout uniformitarian and have been interpreted in various ways. Some have suggested that meteorites impacted the earth, producing a dust cloud which caused a global winter, and killed many forms of life including the dinosaurs. The date established for this, so-called, Cretaceous/Tertiary boundary has been set at about 65 million years ago. See Figure D.2 for a picture of an outcrop containing this boundary. For reasons to be justified later in this report, I will assume this boundary occurs at the end of the global Flood described in Genesis. Rather than 65 million years before present assigned to this boundary, a date of 5500 years will be assumed. This latter date will also be justified later. Sediments, rock layers, and other geologic formations prior to the Cretaceous/Tertiary boundary will all be assumed to have been formed during or prior to the Flood.

4300

APPENDIX E

UNIFORMITARIAN TEMPERATURE CHANGE SINCE THE CRETACEOUS

Cretaceous Warmth

The period from the mid-Cretaceous to the early Tertiary has been calculated to be one of the warmest times in the late Phanerozoic, according to the old-earth position. The average global temperature was probably about 6°C higher than that of today (Barron, 1983), keeping polar regions free of permanent ice. Temperatures were high enough to allow forests and vertebrates to live near both poles, particularly during the peak of warmth in the early Eocene.

The mid-Cretaceous was a time of globally high sea levels and extensive areas of shallow shelf seas, favoring moderate climates and increased evaporation and precipitation (Arthur, Dean, and Schlanger, 1985). It is likely that life on the ocean floor and in the sea was very luxurious during the Cretaceous warmth. Deposition of large quantities of organic-rich sediments in anoxic bottom waters occurred during the mid-Cretaceous but black shales diminished toward the end of the Cretaceous. Sea levels reached their peak in the late Cretaceous and then gradually dropped (Haq et al., 1987). During the late Cretaceous the seas withdrew, exposing larger continental areas and leading to the establishment of more seasonal, continental climates (Hays and Pittman, 1973).

The young-earth Flood model suggests that warm climates and the lack of permanent ice in the polar regions occurred before the Flood. Most, if not all, of the sedimentary layers were produced by the Flood. At some level high in the geologic column, possibly at or near the end of the Cretaceous period, the Flood waters receded, accumulation of sediments decreased, and the climate cooled.

Tertiary Cooling

The climate of the Earth gradually cooled from the late Cretaceous throughout the Tertiary period. Important changes which occurred during this phase include the enhancement of latitudinal climatic zonation and the development of a thermally stratified ocean. Tertiary cooling is not recorded simply by the presence of ancient glacial deposits. In fact, any direct evidence of extensive glacial ice at the poles during the Tertiary is scarce, principally because the rocks in these regions are now covered with ice. However, the presence of ice-rafted debris in deep ocean cores provides positive evidence for the presence of at least seasonal ice.

The principal evidence for Tertiary cooling is documented in the oxygen isotope record of calcareous forams from the ocean. This record indicates a decline in ocean temperatures together with a build-up of ice at the poles (mainly the South Pole). Fig. E.1 shows the isotopic paleo-temperature data for Tertiary planktic forams (open symbols) and benthic forams (closed symbols) near the tropics, primarily from the North Pacific and a few points from the South Atlantic and South Pacific (Douglas and Savin, 1971, 1973, 1975, and unpublished; Savin, et al., 1975). The "Modern" and "Tertiary" benthic temperature scales were calculated assuming water $\delta^{18}O$ values of -0.08‰ and -1.00‰, respectively. To estimate isotopic temperatures of planktic forams add approximately 2.5°C to the appropriate benthic temperature scale.

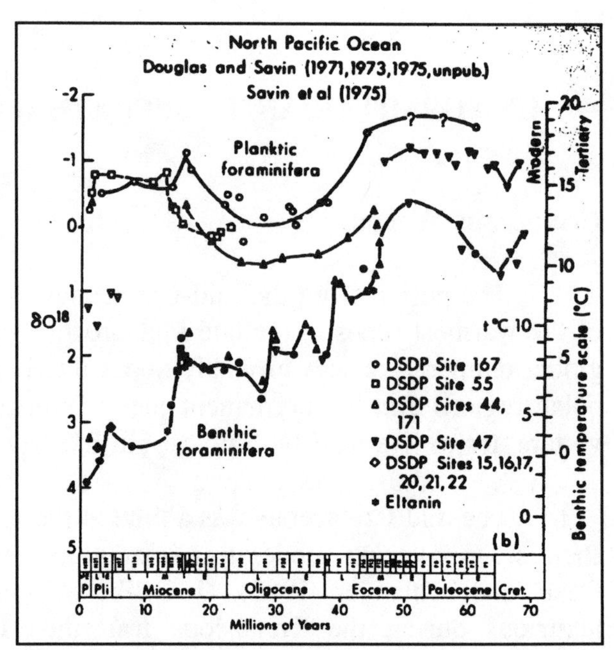

Fig. E.1 *Isotopic data for Tertiary planktic (open symbols) and benthic (closed symbols) forams primarily from the North Pacific. For planktic temperatures add 2.5°C.*

Fig. E.2 shows the isotopic paleo-temperature analyses of planktic and benthic forams from the subAntarctic Pacific (Shackleton and Kennett, 1975). Temperature scales for benthic forams were computed in the same manner as Fig. E.1. However, because of the extreme oceanographic changes in the area during Tertiary time a separate temperature scale for planktic forams was not estimated.

Since planktic forams provide estimates of ocean-surface temperatures and benthic forams provide estimates of ocean-bottom temperatures, Figs. E.1 and E.2 give a general picture of the global changes in ocean temperature during the Tertiary period. The ocean-surface temperature near the tropics has only cooled about 2°C from the Cretaceous to the present, while the ocean-bottom temperature has cooled about 15°C. In the polar region near Antarctica both the ocean-surface and ocean-bottom temperatures have cooled 12 to 18°C. The cold water at the bottom of the tropical oceans is the result of cooling of surface waters in the polar regions which sink and move along the ocean bottom toward the equator. These vertical and horizontal motions induced by buoyancy forces from cooling of the ocean produce, in turn, horizontal coriolis-induced circulations in the

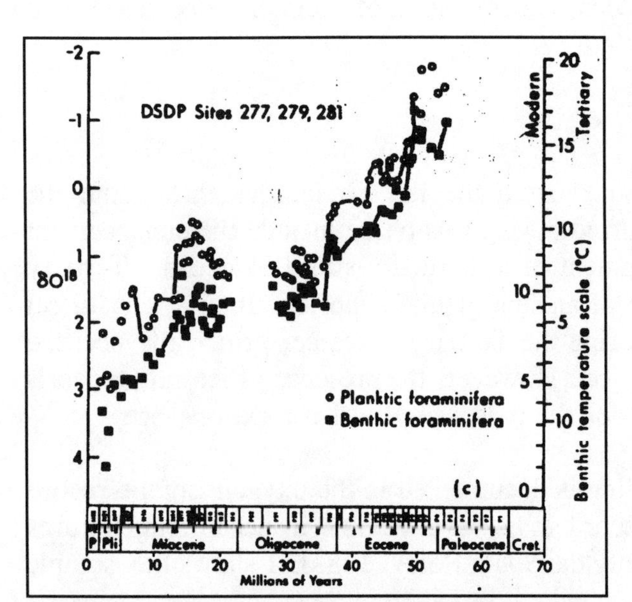

Fig. E.2 *Isotopic data from the subAntarctic Pacific. For planktic temperatures add 2.5°C.*

various ocean basins. See Fig. E.3 for a diagram of today's horizontal circulations in the Pacific Ocean.

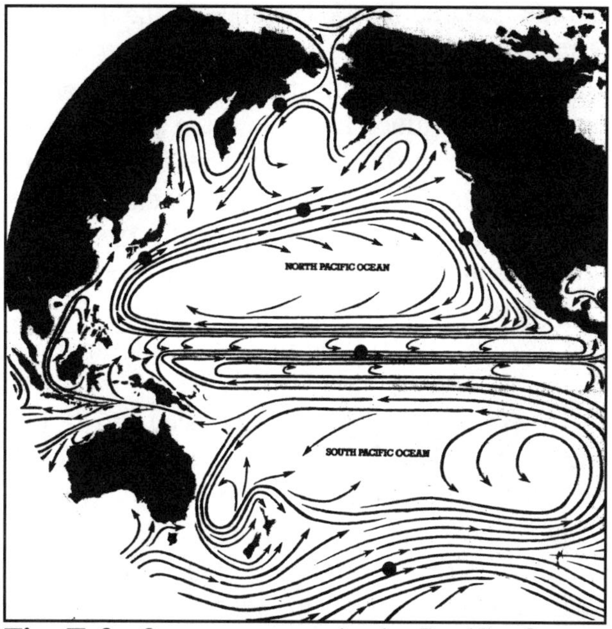

Climate evidence from fossil plant assemblages reflects the same cooling trend. Documentation of the Tertiary cooling is important because it illustrates the crucial transformation phase from a non-glacial to glacial state, as recorded by several geological parameters.

Superimposed on this major cooling phase are several shorter intervals of change, such as the sharp decline in temperatures followed by warming at the Eocene-Oligocene boundary and warmth during the Middle Miocene, followed by cooling. The warming phases are best documented by the spread of temperate vegetation to high latitudes.

Fig. E.3 *Ocean currents in the Pacific Ocean. Two large gyres occur in the North Pacific (clockwise) and the South Pacific (counterclockwise) separated by an equatorial countercurrent.*

Detailed oxygen isotope records are available for most of the Tertiary and have provided a good record of changing ocean temperatures and changing salinities induced by the build-up of continental ice. In general, the isotope record is similar for all major oceans, including the Southern Ocean (Shackleton and Kennett, 1975), South Atlantic (Williams et al., 1983; Poore and Matthews, 1984), North Pacific and Equatorial Pacific (Douglas and Savin, 1973; 1975) and the Indian Ocean (Oberhansli, 1986). Isotope information is summarized in Miller, Fairbanks and Mountain, (1987).

Following low $\delta^{18}O$ values in the late paleocene and early Eocene, both surface and bottom waters had progressively higher values of $\delta^{18}O$ for the rest of the Eocene (note the reversed $\delta^{18}O$ and time scales in Fig. E.2). Since there is little geological evidence for significant quantities of polar ice in the Eocene, the isotopic change has been interpreted as evidence for cooling of ocean waters, rather than for a build-up of ice (Oberhansli and Hsu, 1986). Early Eocene isotope ratios indicate bottom water temperatures of 11 to 15°C. If the isotopic value for Eocene sea water is considered to be the same as that for glacial times (Matthews and Poore, 1980) a temperature of 15°C can be assumed. For an ice-free ocean the corresponding estimate is 11°C. In either case, these temperatures infer that high-latitude surface waters would be too warm for ice formation. Bottom waters in mid-latitudes may have been warmed due to the sinking of warm, highly saline surface water. Planktic forams have indicated high-latitude Eocene ocean surface temperatures up to the 17 to 21°C range.

During the middle Eocene $\delta^{18}O$ values increased with time in two steps, at the early to middle Eocene boundary and in the late middle Eocene. By the end of the Eocene the temperature of bottom waters was about 6 to 10°C and high-latitude oceans are thought to have been ice free. The most significant increase in $\delta^{18}O$ values occurred at the Eocene-Oligocene boundary. Values fell below the threshold of 1.8‰, which implies that the temperature of high-latitude surface waters was cold enough for ice formation. The temperature drop is synchronous in both benthic and planktic forams, indicating the growth of ice, rather than simply a temperature decline (Miller et al., 1987). The presence of ice has also been inferred from a synchronous isotope decrease in the late Oligocene and in the North Atlantic at the Oligocene-Miocene boundary. Apart from these two intervals the Oligocene isotope record is one of only slightly declining ocean temperatures.

The next major cooling occurred in the middle Miocene. A synchronous drop in both benthic and planktic isotopes records the intensification of polar glaciation. The pole-to-equator, surface-water temperature gradient increased during the middle Miocene from 6 to 12°C. High-latitude surface waters cooled significantly but low-latitude waters warmed only slightly (Kennett, 1977). Increasing $\delta^{18}O$ values for equatorial surface waters have been interpreted as indicating a slight drop of ocean temperatures to 22 to 24°C at some time during the mid-Miocene (Shackleton, 1968), although this unusual cooling is not apparent if the isotope values are calculated on the basis of a glaciated ocean composition (Matthews and Poore, 1980).

After a short warming phase in the early Late Miocene a terminal Miocene cooling occurred, causing a major phase of ice growth and intensification of glaciation. Global sea levels dropped markedly. Oxygen isotope values continued to decrease during the early Pliocene. The most depleted $\delta^{18}O$ values over this time are not significantly different from modern values, suggesting similar temperature and ice volume to that of today. During the middle Pliocene, $\delta^{18}O$ ratios reflect an ice-volume increase and/or cooling of bottom waters. When the ice-volume on the earth is increased (enrichment), the heavier isotope (^{18}O) is left behind in sea-water, creating a greater $\delta^{18}O$. The enrichment in the middle Pliocene was a temporary feature in most planktonic and benthic records (Prell, 1984; Keigwin, 1986) but is permanent in a deep-water site (**DSDP** 518) underlain by Antarctic bottom water (Hodell et al., 1985). Prell (1984) suggested that the middle Pliocene event was a brief, temporary increase in ice volume followed by a permanent cooling of bottom water. During the late Pliocene a step-like increase in $\delta^{18}O$ has been interpreted as due to the onset of northern hemisphere glaciation (Shackleton et al., 1984; Backman, 1979). The extent of the influx of ice-rafted debris to the North Atlantic and the range of $\delta^{18}O$ in the late Pliocene are similar to those observed at maxima during the middle Pleistocene (Shackleton et al., 1984; Keigwin, 1986).

The transition from a warm early Pliocene to a cooler middle Pliocene is not reflected in the planktic oxygen-isotope record. Assuming that the $\delta^{18}O$ value of the middle Pliocene was more positive by 0.4‰ than in the early Pliocene as a result of an increase in ice volume, the contrasting planktonic values suggest a 2°C warming of South Atlantic surface waters (to 30° latitude) at the time of ice sheet growth (Keigwin, 1979; Vergnaud and Grazzini, 1980; Weissert et al., 1984).

Quaternary temperature fluctuations

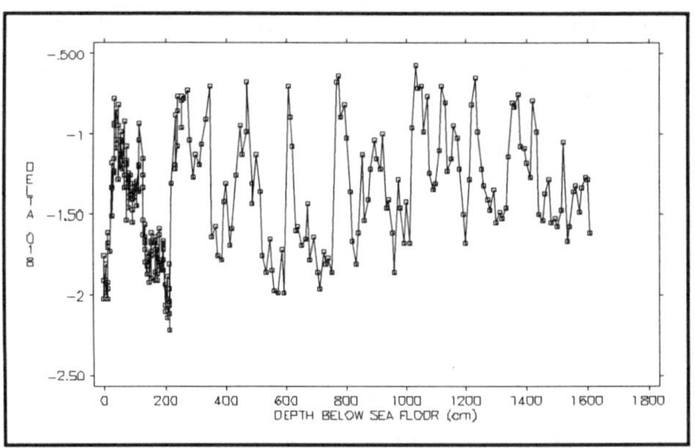

Fig. E.4 *High-resolution isotopic data from the upper 16 meters of sediment at site V28-238 in the equatorial Pacific.*

In contrast to the general cooling trend indicated by $\delta^{18}O$ values from the Cretaceous through the Tertiary, temperatures estimated from $\delta^{18}O$ measurements in the Quaternary period seem to have stopped declining, and have been oscillating around a relatively stable value. Samples in soft sediments of the upper few tens of meters of the ocean floor show small variations in $\delta^{18}O$. Fig. E.4 shows $\delta^{18}O$ as a function of depth for V28-238, a site in the northwest Pacific Ocean near the equator. This particular sediment core was used as a reference core in both the **CLIMAP** and **SPECMAP** Projects. According to the Age Model used in **SPECMAP**, the age of the sediments at the bottom of the core at 16 meters is about 800,000 years.

Note the upper-most peak in $\delta^{18}O$ between 0 and 2 meters in depth. This saw-toothed fluctuation in $\delta^{18}O$ has been interpreted to be associated with the most recent "Ice Age." As ice accumulated on the polar regions, a larger proportion of ^{16}O was removed from the oceans, leaving them enriched with ^{18}O. Forams thus incorporated higher concentrations of ^{18}O into their shells, producing sediments with higher values of $\delta^{18}O$. Glaciation (growth of ice sheets) occurred more slowly than deglaciation (melting of ice sheets), producing a saw-toothed shape to the $\delta^{18}O$ distribution in the sediment as a function of time. The upper-most peak is estimated to have lasted about 100,000 years. Also note other peaks with somewhat similar shape deeper in the sediment. An inverse saw-toothed function is seen in $\delta^{18}O$ of ice sheets. A young-earth, creationist model of ice sheet formation has been developed by Vardiman (1993).

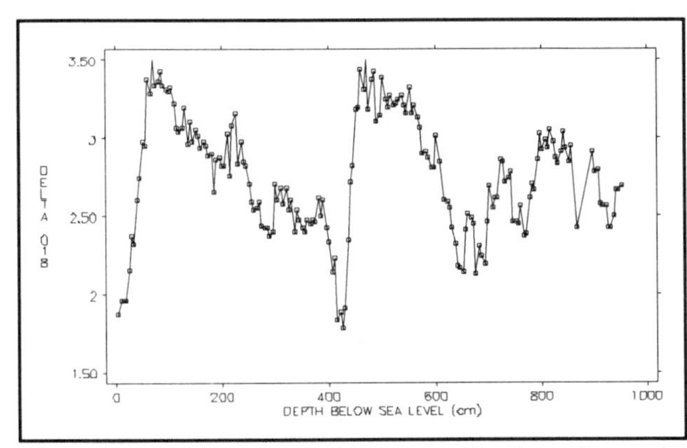

Fig. E.5 *High-resolution isotopic data from the upper 10 meters of sediment at site RC11-120 in the subAntarctic Pacific.*

Records from many cores taken throughout the world show closely similar trends, similar periodicities in $\delta^{18}O$, and presumed ice-volume fluctuations (Martinson, et al., 1987), and similar sea surface temperatures as reconstructed from foram assemblages (**CLIMAP**, 1976, 1981; Hays et al., 1976). For example, Figure E.5 shows the variation of $\delta^{18}O$ as a function of depth for RC11-120, a site in the subAntarctic. The oxygen isotope record is a powerful stratigraphic tool that has provided the basis for a standard climatic model. Layers of sediment have been dated radiometrically, biostratigraphically, and by paleomagnetism. Fluctuations in solar heating due to precession, obliquity, and eccentricity of the earth's orbit around the sun have been computed over the same time interval. A spectral analysis conducted on the $\delta^{18}O$ record found peaks at 19, 23, 41, and 100 thousand years. These peaks have been compared with the solar heating variations due to precession, obliquity, and eccentricity of the earth's orbit. The strongest peak in the $\delta^{18}O$ record occurred at 100,000 years which corresponds with the period of the eccentricity. Fig. E.6 shows the earth in its orbit around the sun and the sources of fluctuation in solar heating.

According to the astronomical theory of palaeoclimates, called the Milankovitch theory by some (Milankovitch, 1930), the long-term variations in the geometry of the earth's orbit are the fundamental cause ("pacemaker") of the succession of Pleistocene ice ages inferred from glacial debris and sea-floor sediments. Spectral analysis of climatic records of the past is believed by some to have provided substantial evidence that the climatic variance is driven in some way by insolation changes caused by orbital forcing. The variable elements of the earth's orbit influence the geographical distribution of solar radiation. However, the total energy flux (insolation) as integrated over the globe annually, depends only upon eccentricity. The total insolation changes due to eccentricity vary by less than 0.2%, and a major task therefore has been to show that this small insolation change could lead to cyclic glaciation of the earth.

Many other fundamental questions remain to be answered. For example, how can such small-amplitude, periodic forcing control the amplitude and sequence of climate events in a complex, non-linear oscillatory system, and, is there a good physical explanation for climate change? The astronomical theory holds that northern hemisphere glaciers grow when summers receive less insolation and melt when they receive more. However, the climate state of the earth is not a simple function of solar insolation. In fact, the earth modulates insolation forcing by reflection, retention, and redistribution of received energy. The terrestrial mechanisms and feedbacks are the major unknowns in predicting climate changes. At present there is no adequate model for climatic change on any time scale.

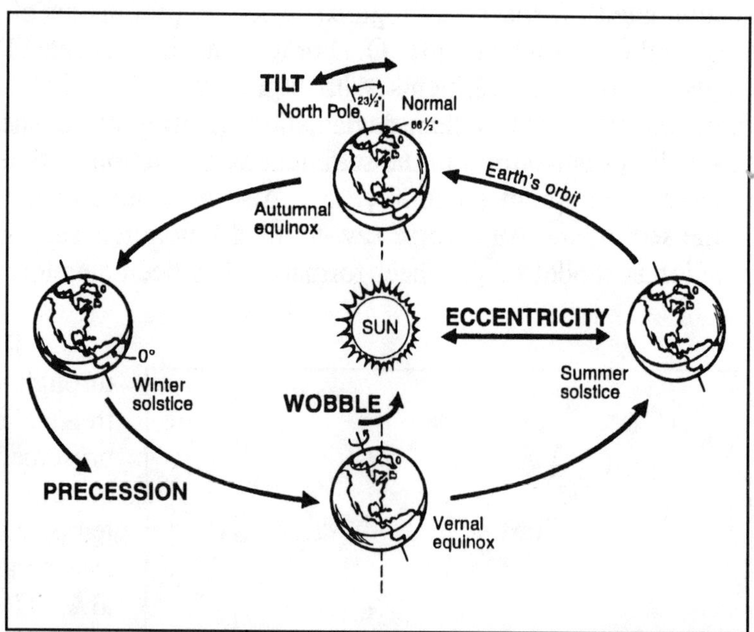

Fig. E.6 *Long-period variations in the orbital parameters of the earth. The axis of the earth wobbles and changes its obliquity. The elliptical shape of the orbit precesses and changes its eccentricity.*

GLOSSARY

Abyssal plain--a flat region of the ocean floor, usually at the base of the *continental rise*. It is formed by the deposition of turbidity-current and pelagic sediments that obscure the preexisting topography.

Accumulation rate--the rate at which sediment accumulates on the ocean floor. In units of meters per year or centimeters per 1000 years.

Advanced piston corer--one of the drilling tools used to recover undisturbed cores in soft ooze and sediments at the seafloor.

Age profile--the age assigned to each layer of sediment along a core recovered from the seafloor. This age can vary dramatically depending upon the assumptions about accumulation rates and processes.

Algae--photosynthetic aquatic plants, including both seaweeds and their fresh-water allies. They range in size from simple unicellular forms to giant kelps several meters long.

Anoxic--a condition under which little or no oxygen is available for organic processes.

Antediluvians--people who lived before the Genesis Flood.

Aragonite--a white, yellowish, or gray orthorhombic form of $CaCO_3$. Aragonite has a greater density and hardness, and a less distinct cleavage, than *calcite*, and is also less stable and less common. It occurs in fibrous aggregates in beds of gypsum and iron ore; as a deposit from hot springs; and as a major constituent of shallow marine *muds* and the upper parts of *coral reefs*. Aragonite is also an important constituent of the pearl and of some shells.

Archive core--half of each core which is stored at one of the three repositories in New York, Texas, or California and is never sampled. It can be studied by non-destructive means.

Asteroid--one of many small celestial bodies in orbit around the sun. Most asteroid orbits are between those of Mars and Jupiter. When an asteroid or fragment has fallen to earth it is called a meteorite.

Astronomical Theory--the theory that "Ice Ages" are long-period cyclical phenomena initiated by harmonic changes in the orbital parameters of the earth-sun system. It was first suggested by Milutin Milankovitch in 1930 and has been refined and generally accepted in the late 20th century.

Atmosphere--the gaseous envelope surrounding a planet. Specifically for the Earth, the envelope of air composed of approximately 78% nitrogen, 21% oxygen, and 1% other minor gases.

Bed--the smallest formal unit in the hierarchy of stratigraphic units which is distinguishable from layers above and below. A bed commonly ranges in thickness from a centimeter to a few meters.

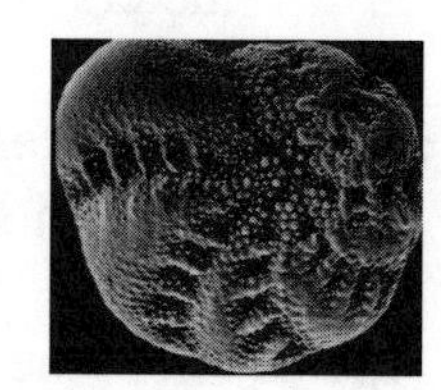

Benthic foram--a single celled organism with no tissue or organs which lives near or on the seafloor characterized by the presence of a small shell (test) of one or many chambers composed of calcite or silica. Most forams are marine but freshwater forms are known.

Biogenous--formed directly by the physiological activity of organisms, either plant or animal. For example, the formation of biogenous sediments by the death and sedimentation of forams.

Biological potential--the possibility of organic growth due to the presence of adequate nutrients in the sea.

Biosphere--all the area occupied or favorable for occupation by living organisms in the lithosphere, hydrosphere, and atmosphere.

Black shale--a dark, thinly laminated carbonaceous shale, exceptionally rich in organic matter (5% or more carbon content) and sulfide and often containing unusual concentrations of certain trace elements.

Bloom--an aquatic growth of algae in such concentration as to cause discoloration of the water.

Boundary conditions--the state of a physical system at specific locations and times used to derive analytic solutions. Often the boundary conditions are specified at the initial time and outer limits of the system.

Bulk sample--a sample of sediment processed for a particular isotope without separation by organism, species, or other division.

Calcareous constituent--part of a sample or substance which contains calcium carbonate. When applied to a rock name it implies that as much as 50% of the rock is calcium carbonate.

Calcareous nannofossil--extremely small marine (usually algal) fossils composed of calcium carbonate shells (tests). The fossils are smaller than microfossils and are discoasters and coccoliths.

Calcite--a common rock-forming mineral: $CaCO_3$. Calcite is usually white, colorless, or pale shades of gray, yellow, and blue. It is the principle constituent of limestone.

Calcium carbonate--a solid chemical compound composed of the elements calcium, carbon, and oxygen with the formula $CaCO_3$.

Calcium carbonate compensation depth--the level in the ocean below which the rate of solution of calcium carbonate exceeds the rate of deposition. This level is typically 4,000 to 5,000 meters below sea level.

Cambrian period--the earliest period of the Paleozoic era, thought to have covered the time span between 500 and 570 million years ago by the evolutionary time scale. Named after Cambria, the Roman name for Wales, where rocks of this type were first studied.

Carbon dioxide--a gaseous chemical compound composed of the elements carbon and oxygen with the formula CO_2.

Carboniferous period--the Mississippian and Pennsylvanian periods of the Paleozoic era combined, ranging from about 280 to 345 million years ago by the evolutionary time scale. Rocks in this period are widespread and contain much coal.

Catastrophic--sudden violent, short-lived, more or less worldwide events in Earth history outside our present experience or knowledge.

Catastrophic process--the mechanism or the method by which sudden violent, short-lived, more or less worldwide events in earth history outside our present experience or knowledge occurred.

Cementation--the process by which coarse sediments become *lithified* or consolidated into hard, compact rocks, usually through *deposition* or *precipitation* of minerals in the spaces among the individual grains of the sediment.

Cenozoic Era--an era of geologic time, from the beginning of the Tertiary period to the present. The Cenozoic began about 65 million years ago according to the evolutionary time scale. It is characterized by the abundance of mammals, mollusks, birds, and angiosperms.

Chronology--the science of dating events accurately and arranging them in the order of occurrence.

Chronostratigraphy--the science of dating strata (layers of sediment) accurately and arranging them in the order of occurrence.

Clay--a loose, earthy, extremely fine-grained, natural sediment or soft rock composed primarily of clay-size or colloidal particles and characterized by high plasticity and by a considerable content of clay minerals.

Clay minerals--a complex and loosely-defined group of finely crystalline, metacolloidal, or amorphous hydrous silicates, essentially aluminum and sometime of magnesium and iron.

CLIMAP--(Climate/Long-Range Investigation, Mapping, and Prediction) a concentrated effort begun in the early 1970s to synthesize the work of past generations of geologists, add new information from the rapidly expanding area of Pleistocene oceanography, and reconstruct the climate of the earth's surface at the mid-point of the "last" glacial maximum (18,000 years before the present according to the evolutionary time scale).

Climatic zonation--the tendency to form regions or zones of similar climate due to similar latitude, altitude, and distance from the ocean.

Coccolith--a general term applied to various microscopic calcareous structural elements or button-like plates derived from mostly marine planktonic algae.

Coccolithophore--any of numerous minute mostly marine planktonic algae.

Continental climate--the weather of a region near the center of a continent where little oceanic influence is felt characterized by seasonal temperature extremes.

Continental crust--that type of earth's *crust* which underlies the continents and the *continental shelves*; it is equivalent to the *sial*, and ranges in thickness from about 35 kilometers to as much as 60 kilometers under mountain ranges.

Continental ice--ice of considerable thickness completely covering a large part of a continent, such as in the ice sheets covering Antarctica and Greenland.

Continental ice sheet--a layer of ice of considerable thickness completely covering a large part of a continent, such as in the ice sheets covering Antarctica and Greenland.

Continental margin--the ocean floor that is between the shoreline and the *abyssal plain*.

Continental rise--that part of the *continental margin* that is between the *continental slope* and the *abyssal plain.*

Continental shelf--that part of the *continental margin* that is between the shoreline and the *continental slope*.

Continental slope--that part of the *continental margin* that is between the *continental shelf* and the *continental rise* characterized by a relatively steep slope.

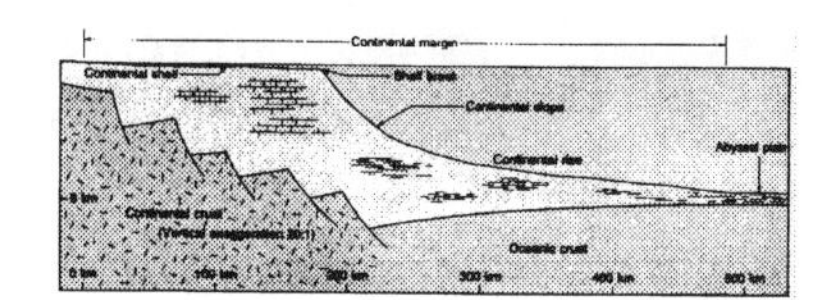

Cooling curve--the relationship between temperature and time plotted on a two-dimensional graph.

Coral reefs--a mound or ridge of in-place coral colonies and accumulated skeletal fragments, carbonate sand, and limestone resulting from organic secretion of calcium carbonate that lithifies colonies and sands. A coral reef is built up around a potentially wave- and surf-resistant framework, especially of coral colonies but often including many algae; the framework may constitute less than half of the reef volume. Coral reefs occur today throughout the tropics, wherever the temperature is suitable (generally above 18°C, a winter minimum).

Core depths--the distance downward along a core from the ocean floor.

Core--the central zone or nucleus of the earth's interior. It is solid, having a radius of approximately 1300 kilometers.

Cores--a cylindrical section of rock or sediment, usually 5-10 centimeters in diameter and up to several meters in length, penetrated by a core bit, and brought to the surface for geologic examination or laboratory analysis.

Cork--a borehole seal that allows instruments to be placed down the hole to monitor measurements after the core has been removed, including temperature and pressure.

Creationist model--a conceptual or analytic representation of earth processes within a Biblical framework.

Cretaceous Period--the final period of the Mesozoic era and before the Tertiary period of the Cenozoic era, ranging from between 65 and 135 million years ago by the evolutionary scale. It is named after the Latin word for chalk ("creta") because of the English chalk beds in this strata.

Cretaceous/Tertiary boundary--the interface between the Cretaceous and Tertiary strata in the geologic column. A sharp unconformity occurs in many places around the world at this level, often containing high concentrations of unique element iridium, thought by some to be associated with catastrophic impacts of meteorites.

Crust--the outermost layer or shell of the earth above the *Mohorovicic discontinuity*. It represents less than 0.1% of the earth's total volume.

Cryogenic magnetometer--an instrument used on board drilling ships to determine the magnitude and direction of residual magnetic fields in core samples. Extremely cold refrigerants are used to increase the sensitivity of the instrument.

Cryosphere--the portion of the earth's surface that is perennially frozen.

Deep Sea Drilling Project (DSDP)--an internationally sponsored scientific project to drill cores in the sea floor, interpret the results, and report in a standardized format. DSDP completed corings at 624 sites around the world during the period from 1968 to 1983.

$\delta^{13}C$--the relative difference in the ratio of $^{13}C/^{12}C$ between a given sample and a reference sample (normally from the Peedee Belemnite Formation in South Carolina) expressed in parts per thousand.

$\delta^{18}O$--the relative difference in the ratio of $^{18}O/^{16}O$ between a given sample and a reference sample (normally from the Peedee Belemnite Formation in South Carolina) expressed in parts per thousand.

Deluge--the global, catastrophic Flood described in Genesis which produced the fossiliferous sedimentary rocks a few thousand years ago.

Density--having the quality of being packed tightly together. The measure of mass per unit volume (gm/cm^3).

Deposition--the process whereby sediment is accumulated on the sea floor.

Depositional basin--a region on the sea floor where similar sediments are deposited, typically of a relatively lower and similar elevation than the surroundings.

Depositional temperature--the temperature of the water in which the organisms lived before being deposited as sediment after their death.

Detrital origin--the source of mineral grains derived from the mechanical disintegration of parent rock.

Devonian period--a period of the Paleozoic era, ranging from about 345 to 400 million years ago, according to the evolutionary time scale. Named after Devonshire, England where this system of rocks containing many fossil fishes was first studied.

Diagenesis--all the chemical, physical, and biologic changes undergone by a sediment after its initial deposition, and during and after its lithification, exclusive of weathering and metamorphism.

Diamond coring system--a type of bit and mechanism for extracting core in hard rock. It contains diamonds imbedded in the bit to cut into the rock.

Diatom--a microscopic, single-celled plant which grows in both marine and fresh water. Diatoms secrete walls of silica in a great variety of forms which may accumulate in sediments in enormous numbers.

Dilution--to thin down or weaken by mixing with something else.

Discoaster--tiny star- or rosette-shaped calcareous plates, 10-35 microns in diameter, generally believed to be the remains of a planktonic organism.

Discrete sediment--sediment composed of weathered and unconsolidated material from the surface of the land.

Dissolution--a process of chemical weathering by which mineral and rock material passed into solution; e.g. removal of the calcium carbonate in limestone by carbonic acid derived from rainwater.

Drill collar--a length of extra-heavy, thick-walled drill pipe in a rotary drill string directly above either the bit or the core barrel, to concentrate weight and give rigidity so that the bit will cut properly.

Drilling mud--a carefully formulated heavy suspension, usually in water but sometimes in oil, used in rotary drilling. It commonly consists of bentonitic clays, chemical additives, and weighing materials such as barite. It is pumped continuously down the drill pipe, out through openings in the drill bit, and back up in the annulus between the pipe and the walls of the hole to a surface pit where it is screened and reintroduced through the mud pump. The mud is used to lubricate and cool the bit; to carry the cuttings up from the bottom; and to prevent blowouts and cave-ins by plastering friable or porous formations with mud cake, and maintaining a hydrostatic pressure in the borehole offsetting pressures of fluids that may exist in the formation.

Eccentricity--deviation from a circular shape. In elliptic orbits, it is the ratio of the distance between foci to the major axis of the orbit.

Ecliptic--the plane of the earth's orbit around the sun.

Ecology--the branch of biology that deals with the relationships between living organisms and their environment.

Energy flux--the rate of energy flow in units of energy per unit volume times the flow velocity.

Enrichment--a process of mineral deposition whereby near-surface oxidation produces acidic solutions that leach metals, carry them downward, and reprecipitate them, thus enriching sulfide minerals already present.

Eocene epoch--an epoch of the Tertiary period, after the Paleocene and before the Oligocene.

Epoch--a geologic time unit longer than an age and shorter than a period. Also a term used informally to designate a length (usually short) of geologic time.

Equatorial counter current--the return flow of eastward-moving surface water in the equatorial oceans in a narrow band between westward-moving flow to the north and south.

Equinox--either of the two points of intersection of the sun's apparent annual path and the plane of the Earth's equator. Popularly, the time at which the sun passes directly above the equator; the "time" of the equinox. When day and night have the same (equal) length.

Era--a geologic time unit next in order of magnitude below an eon; e.g., the Paleozoic Era, the Mesozoic Era, and the Cenozoic Era. Each of these eras includes two or more periods, during each of which a system of rocks was formed.

Evaporation--the change in phase from liquid to vapor; e.g. the change from liquid water to water vapor by the absorption of heat.

Evolutionary model--a conceptual or analytic representation of earth processes according to a uniformitarian processes over billions of years.

Evolutionary time scale--the long time scale assigned to the geologic column and development of life on earth required by the evolutionary model.

Exponential decrease--the reduction in some quantity according to a rapidly decreasing power function with a base $e = 2.71828...$ This function is frequently observed in geophysics when a sudden change in boundary conditions expressed in differential equations occurs.

Flood/postflood boundary--a hypothetical location in the geologic column where the events near the end of the Genesis Flood occurred. Although reference is often made to a boundary, it is more likely that a zone would be expected since the intensity and type of Flood processes probably did not change suddenly.

Foram--an order of microscopic protozoa characterized by the presence of a shell (test) of one to many chambers composed of secreted calcite (rarely silica or aragonite). Most forams are marine but freshwater forms are known. They range from the Cambrian to the present. Actually, "foram" is a popular term for the more technical name "foraminifer" (plural "foraminifera").

Foram assemblages--a group of foram fossils that occur at the same stratigraphic level; often with a connotation also of localized geographic extent. For its accurate interpretation, an assemblage should be relatively homogeneous or uniformly heterogeneous.

Gametogenesis--the uniting of one reproductive cell with another to form a zygote which can develop into a new individual.

Geochemistry--the study of the distribution and amounts of chemical elements in minerals, ores, rocks, soils, water, and the atmosphere, and the study of their circulation in nature.

Geologic column--a composite diagram that shows in a single column the subdivisions of part or all of the sequence of stratigraphic units of a given locality or region with the oldest at the bottom and the youngest at the top.

Geologic strata--sheetlike bodies or layers of sedimentary rock, visually separable from other layers above and below. Strata have been defined as stratigraphic units that may be composed of a number of *beds*.

Geologic time scale--an arbitrary chronologic arrangement or sequence of geologic events, used as a measure of the relative or absolute duration or age of any part of geologic time, and usually presented in the form of a chart showing the names of the various geologic time units. The time scale can be expressed according to an evolutionary model or a creationist model.

Geological Society of America (GSA)--a group of geologists, primarily in North America, organized to encourage research, education, and publication in geology. This society is a daughter organization of the first such society in England called the **Geological Society**.

Geologist--a specialist in the science dealing with the structure of the earth's crust and the formation and development of its various layers.

Geophysicist--a specialist in the sciences dealing with the physics of the earth including the atmospheric sciences, ocean sciences, planetology, seismology, tectonophysics, volcanology, etc.

Glacial--of or relating to the presence and activities of ice, glaciers, and/or the "Ice Age".

Glacial epoch--any part of geologic time, from Precambrian onward, according to the evolutionary time scale, in which the climate was notably cold in both the northern and southern hemispheres, and widespread glaciers moved toward the equator and covered a much larger total area than those of the present day.

Glacial-marine sediment--sediments on the sea floor which contain glacial material.

Global flood--the catastrophic Flood described in Genesis which covered all the mountains over the entire earth and produced the fossiliferous sedimentary rocks a few thousand years ago.

Global winter--the catastrophic cooling of the entire earth resulting in the extinction of life as we know it, following a nuclear exchange between Russia and the United States. Carl Sagan hypothesized the scenario in which a nuclear exchange would ignite such large fires and loft so much dust into the upper atmosphere that solar heating would be decreased dramatically for many years causing catastrophic cooling.

Glomar Challenger--the primary research ship used during DSDP to drill sea floor sediment cores around the Earth.

Graded bedding--a type of bedding in which each layer displays a gradual and progressive change in particle size, usually from coarse at the base of the *bed* to fine at the top. It may form under conditions in which the velocity of the prevailing current declined in a gradual manner, as by *deposition* from a single, short-lived *turbidity current*.

GRAPE (Gamma Ray Attenuation Porosity Evaluator)--an instrument and method used in DSDP and ODP and other laboratories to measure bulk density and porosity of a sediment sample. The instrument uses measured values of the attenuation of gamma rays through a sample and assumed constants for sea water.

Habitat--the particular environment or place where an organism or species tends to live.

Hard rock--a term used loosely for *igneous* or *metamorphic rock*, as distinguished from *sedimentary rock*.

Hard rock guide base--a funnel-shaped apparatus placed on the sea floor to guide the drill bit to a specific site and hold it in place until the drill makes enough penetration into igneous or metamorphic rock to maintain its own position.

Hole--the cavity left after a core has been removed from a drill site.

Holocene epoch--the most recent epoch of the Quaternary period, ranging from about 8,000 years ago to the present, according to the evolutionary time scale.

Hydrocarbons--organic compounds in gaseous, liquid, or solid states, consisting solely of carbon and hydrogen.

Hydrogenous--containing high moisture contents; e.g. brown coal.

Hydrosphere--the waters of the earth, as distinguished from the rocks (lithosphere), living things (biosphere), and the air (atmosphere).

Ice Age--a loosely used synonym of *glacial epoch*, or time of glacial activity. Specifically, the latest of many *glacial epochs* which last about 100,000 years each separated by shorter interglacials, according to the evolutionary model. The creationist model of the *Ice Age* proposes a single glacial event which was caused by and occurred after the *Genesis Flood* a few thousand years ago.

Ice core--a cylindrical section of ice, usually 10-20 centimeters in diameter and up to several meters in length, penetrated by a core bit, and brought to the surface for geologic examination or laboratory analysis.

Ice-rafted sediment--sediment which contains materials such as boulders or *till*, which is deposited by the melting of floating ice containing it.

Igneous rock--a rock or mineral that solidified from molten or partly molten material, i.e. from a magma.

Initial reports--portions of the volumes published by DSDP and ODP within about a year after a cruise, reporting on site characteristics, core photographs, and other preliminary data and results.

Inorganic--pertaining or relating to a compound that contains no carbon.

Insolation--the combined direct solar and sky radiation reaching a given body; e.g., the earth. Also, the rate at which it is received in units of energy per unit time per unit area.

Iron oxide--the loosely used name for oxidized iron. It can take various forms; e.g., hematite (Fe_2O_3), magnetite (Fe_3O_4), and others.

Isostatic adjustment--the adjustment of the *lithosphere* of the earth to maintain equilibrium among units of varying mass and density; excess mass above is balanced by a deficit of density below, and vice versa.

Isotope--one of two or more species of the same chemical element, i.e. having the same number of protons in the nucleus, but differing from one another by having a different number of neutrons. The isotopes of an element have slightly different physical and chemical properties, owing to their mass differences, by which they can be separated.

Isotopic composition--the relative quantity of different isotopes of the same element.

Isotopic equilibrium--the condition under which the concentration of various isotopes of an element remains constant.

Isotopic fractionation--the relative enrichment of one isotope of an element over another, owing to slight variations in their physical and chemical properties. It is proportional to differences in their masses and often occurs at phase changes. In the case of the fractionation of oxygen isotopes the relative enrichment can be used as a measure of temperature.

Isotopic temperature--a method of estimating the temperature by using the property of *isotopic fractionation.*

JOIDES (Joint Oceanographic Institutions for Deep Earth Sampling)--an international organization responsible for developing scientific plans and providing general scientific direction for the completed DSDP and the ongoing ODP. Some of the U.S. institutions associated with JOIDES are: University of California, San Diego, Scripps Institute of Oceanography; Columbia University, Lamont-Doherty Earth Observatory; Oregon State University, College of Oceanic and Atmospheric Sciences; Texas A&M University, College of Geosciences and Maritime Studies; University of Washington, College of Ocean and Fishery Sciences; and Woods Hole Oceanographic Institution.

JOIDES Resolution--the primary research ship used during ODP to drill sea floor sediment cores around the earth.

Jurassic period--the second period of the Mesozoic era (after the Triassic and before the Cretaceous), ranging from about 135 to 190 million years ago, according to the evolutionary time scale. Named after the Jura Mountains between France and Switzerland, in which rocks of this period were first studied.

Kerogen--fossilized insoluble organic material found in *sedimentary rocks*, usually shales, which can be converted by distillation to petroleum products.

Leg--a segment of the route that a drilling ship makes during its career. A **leg** typically includes 5-10 drilling sites selected in a particular geographic area to obtain data for a particular scientific purpose; e.g., **Leg** 96 of DSDP drilled at 11 sites in the Gulf of Mexico.

Lithification--the conversion of a newly deposited, unconsolidated sediment into a coherent, solid rock, involving processes such as cementation, compaction, desiccation, and crystallization.

Lithogenous--said of stone-secreting organisms, such as coral polyps.

Lithosphere--the solid portion of the earth, as compared with the *atmosphere* and the *hydrosphere*. It includes the *crust* and the upper *mantle* and is on the order of 100 kilometers in thickness.

Local flood--the condition that occurs when water overflows the natural or artificial confines of a stream or other body of water of limited geographic extent, or accumulates by drainage over low-lying areas. As opposed to *global flood* whereby the entire earth was covered by water.

Local magnetic reversals--a change in the *magnetic field* between normal polarity and reversed polarity in a limited geographic region, hypothesized by the *creation model*. The local change is due to disruption of normal magnetic field lines by motions in magma generating secondary fields.

Longevity--a great span of life.

Magnetic field--the space occupied by magnetic lines of force; e.g., in the vicinity of the Earth, curved magnetic lines enter and exit the magnetic poles of the earth, believed to be created by electrical currents flowing in the earth's *core*.

Magnetic reversals--a change in the entire earth's *magnetic field* between normal polarity and reversed polarity, hypothesized by the *evolutionary model*. The *creation model* suggests magnetic reversals do not occur, but rather, *local magnetic reversals*.

Magnetization--the magnetic moment per unit volume due to induced magnetization and remanent magnetization.

Manganese nodule--a small, irregular, black to brown, laminated concretionary mass consisting primarily of manganese salts and manganese-oxide minerals. These nodules are abundant on the floors of the world's oceans as a result of *pelagic* sedimentation or precipitation, especially in areas of slow *deposition*. Manganese nodules range in size from a few millimeters to 25 centimeters in diameter and have an average weight of 115 grams, although a nodule weighing 770 kilograms has been found.

Mantle--the zone of the earth below the *crust* and above the *core*.

Maritime climate--the climate of islands and of land areas bordering the ocean, characterized by only moderate diurnal and annual temperature ranges and by the occurrence of maximum and minimum temperatures longer after the summer and winter *solstices*, respectively, than in a *continental climate*.

Mass spectrometer--an instrument for producing and measuring a *mass spectrum*. It is especially useful for determining molecular weights and relative abundances of *isotopes* within a compound.

Mass spectrum--the pattern of relative abundances of ions of different atomic or molecular mass within a sample. It frequently refers to the measured relative abundances of various *isotopes* of a given element.

Mesozoic era--an era of geologic time, from the end of the *Paleozoic* to the beginning of the *Cenozoic*, ranging from about 65 to 225 million years ago, according to the *evolutionary time scale*.

Metamorphic rock--any rock derived from pre-existing rocks by mineralological, chemical, and/or structural changes, essentially in the solid state, in response to marked changes in temperature, pressure, shearing stress, and chemical environment, generally at depth in the Earth's *crust*.

Metamorphism--the mineralological, chemical, and structural adjustment of solid rocks to physical and chemical conditions which have generally been imposed at depth in the *crust* below the surface zones of *weathering* and *cementation*, and differ from the conditions under which the rocks in question originated.

Meteorite--any *meteoroid* that has fallen to the earth's surface in one piece or in fragments without being completely vaporized by intense frictional heating during its passage through the *atmosphere*; a stony or metallic meteoroid large enough to survive passage through the Earth's *atmosphere* and reach the ground.

Meteoroid--one of the countless solid objects moving in interplanetary space, distinguished from *asteroids* and planets by its smaller size but considerably larger than an atom or molecule.

Micron--a unit of length going out of use, equal to 1 millionth of a meter. The new name for this unit of length is the micrometer.

Micronodule--a manganese nodule smaller than 1 millimeter.

Microorganism--any microscopic or ultramicroscopic animal or vegetable organism; especially, any of the bacteria, protozoa, viruses, etc.

Milankovitch theory--the theory that slight, long-period changes in orientation and distance from the sun which the earth experiences in its orbit result in small variations in *insolation* which, in turn, trigger "*Ice Ages*".

Miocene epoch--an *epoch* of the upper *Tertiary period*, after the *Oligocene* and before the *Pliocene*.

Mississippian--a period of the Paleozoic era, ranging between 320 and 345 million years ago, according to the evolutionary time scale. It was named after the Mississippi River valley, in which there are good exposures of these rocks.

Model--a conceptual or analytic representation of events and processes of earth history. In an analytic model the representation is made in mathematical formulae which can sometimes be used to simulate historic conditions and predict future events.

Mohorovicic discontinuity--the boundary that separates the earth's crust from the mantle below. It marks the level in the Earth at which P-wave velocities change abruptly from 6.7 - 7.2 kilometers per second (in the lower crust) to 7.6 - 8.6 kilometers per second (at the top of the upper mantle). The discontinuity probably represents a chemical change from basaltic materials above to peridotitic materials below, rather than a phase change.

Mollusks--a solitary marine invertebrate characterized by a nonsegmented shell that is bilaterally symmetric. Included in the mollusks are the gastropods, pelecypids, and cephalopods.

Morphology--the study of shape, dimensions, and distributions; e.g. when applied to biology, the branch of biology that deals with the form and structure of animals and plants or their fossil remains.

Mud--a sticky, fine-grained (less than 62 *micron* particle diameters), marine *sediment*, either *pelagic* or *terrigenous*. Muds are the most common *sediment* covering the deep-ocean bottom, as well as the *continental shelf* and *slope*. They are most often described by color; e.g., blue mud, black mud, red mud, etc.

National Geophysical Data Center (NGDC)--a government-funded data center in Boulder, Colorado, which archives geophysical data and distributes it to interested users for a nominal fee; e.g., ice core data, sea floor sediment data, etc.

National Science Foundation (NSF)--a U.S. government agency which funds basic research; e.g. DSDP and ODP were heavily funded by NSF.

Neogene period--an interval in the geologic time scale incorporating the Miocene and Pliocene of the Tertiary period.

Newton's iterative techniques--mathematical methods for solving transcendental and differential equations by repetitively computing solutions closer to the final solution by using differences and derivatives.

Nutrient--any inorganic or organic compound used to sustain plant life in the ocean; e.g., silica for diatoms or iron for algae.

Obliquity--the angle between the plane of the *ecliptic* (or the plane of the Earth's orbit) and the plane of the earth's equator; the "tilt" of the earth.

Oceanic crust--that type of the earth's *crust* which underlies the ocean basins; it is characterized by the absence of the *sialic* layer (rocks rich in silica and alumina, characteristic of the upper *continental crust*). The oceanic crust is about 5-10 kilometers thick.

Ocean Drilling Project (ODP)--an internationally sponsored scientific project to drill cores in the sea floor, interpret the results, and report in a standardized format. ODP has completed corings on more than 50 legs around the world during the period from 1983 to the present.

Oceanographer--a specialist in the science dealing with the physical, chemical, biologic, and geologic aspects of the earth's oceans.

Oil shale--a *kerogen*-bearing, finely laminated brown or black *sedimentary rock* that will yield liquid or gaseous *hydrocarbons* on distillation.

Old-earth model--a creationist representation of earth processes which assumes that the original materials which compose the earth are ancient. Old-earth models may be either catastrophic or uniformitarian.

Oligocene epoch--an *epoch* of the early *Tertiary* period, after the *Eocene* and before the *Miocene*.

Ooze--a soft, soupy mud or slime, typically found covering the bottom of a river, estuary, or lake. It is a wet, earthy material that flows gently, or that yields easily to pressure. It is sometimes defined as a deposit containing more than 30% biogenous constituents by volume.

Opal--a mineral or mineral gel: SiO_2 Nh_2O. It has a varying proportion of water (as much as 20% but usually 3 to 9%); it occurs in nearly all colors, is transparent to nearly opaque and typically exhibits a marked iridescent *play of color*.

Orbital forcing--the concept in the *Astronomical Theory* that the long-period cycles in the orbital parameters of the earth-sun system produce coincident cycles in heating of the Earth and layering of sediments on the sea floor.

Ordovician period--the second earliest period of the *Paleozoic* era (after the *Cambrian* and before the *Silurian*), ranging from between 440 and 500 million years, according to the *evolutionary time scale*. Named after the Celtic tribe called the Ordovices.

Organic-rich sediment--sediment containing large quantities of organic material, thus containing *hydrocarbons*.

Osmosis--the movement at unequal rates of a solvent through a semipermeable membrane, which usually separates the solvent and a solution, or a dilute solution and a more concentrated one, until the solutions on both sides of the membrane are equally strong.

Oxidation--the process of uniting a compound with oxygen.

Oxygen isotope fractionation--the relative enrichment of oxygen 18 over oxygen 16 in seawater during evaporation, owing to the greater vapor pressure of oxygen 16 and consequently, its greater evaporation rate.

Oxygen isotopes--^{16}O and ^{18}O. ^{18}O has two more neutrons than ^{16}O but the same number of protons.

Paleocene epoch--an *epoch* of the early *Tertiary period* after the *Cretaceous* and before the *Eocene*.

Paleochronometer--a loosely used term referring to a geologic variable which can be used to estimate the time at which events occurred.

Paleoclimate--the climate in the past when conditions were different than today.

Paleogene period--a combination of the *Paleocene*, *Eocene*, and *Oligocene epochs* in the *Tertiary period*.

Paleomagnetics--the study of natural remanent *magnetization* in order to determine the intensity and direction of the earth's *magnetic field* in the geologic past.

Paleontology--the study of life in past geologic time, based on fossil plants and animals and including phylogeny, their relationships to existing plants, animals, and environments, and the chronology of earth's history.

Paleotemperature--the mean temperature at a given time or place in geologic history; especially the paleoclimatic temperature of the sea.

Paleozoic era--an era of geologic time, from the end of the *Precambrian* to the beginning of the *Mesozoic*, ranging from about 225 to 570 million years ago, according to the *evolutionary time scale*.

Parameter--a constant or variable in a mathematical expression that distinguishes various specific situations.

Peedee belemnite--the *calcium carbonate* deposit in the *Peedee formation* of South Carolina used as a standard of comparison in determining the isotopic composition of carbon and oxygen to estimate *paleotemperatures.* The deposit is a collection of fossil shells (tests) from belemnites, which are cephalopods which died during the interval from the *Carboniferous* to *Eocene*. H.C. Urey first used this deposit in 1951 as his standard for determining *paleotemperatures* on the basis of the *fractionation* of oxygen *isotopes.*

Peedee formation--a set of *calcium carbonate* units in the upper *Cretaceous* in South Carolina used as a standard for estimates of *paleotemperature* from oxygen *isotope fractionation.*

Pelagic--said of marine organisms whose environment is the open ocean, rather than the bottom or shore areas. Also marine sediment in which the fraction derived from the continents indicates deposition from a dilute mineral suspension distributed throughout deep-ocean water.

Pennsylvanian--a period of the *Paleozoic era* (after the *Mississippian* and before the *Permian*), ranging from about 280 to 320 million years ago, according to the *evolutionary time scale.* It is named after the state of Pennsylvania in which rocks of this type are widespread and yield much coal.

Period--a geologic time unit longer than an *epoch* and a subdivision of an *era.* It is the fundamental unit of the standard *evolutionary time scale.* Also used informally to designate a length of geologic time; e.g. *glacial period.*

Permian period--the last period of the *Paleozoic era*, ranging from about 225 to 280 million years ago, according to the *evolutionary time scale.* It is named after the province of Perm, USSR, where rocks of this type were first studied.

Petrology--the branch of geology dealing with the origin, occurrence, structure, and history of rocks, especially *igneous* and *metamorphic* rocks.

Phanerozoic--that part of geologic time represented by rocks in which the evidence of life is abundant; i.e., *Cambrian* and later.

Phosphate--a mineral containing tetrahedral PO_4^{-3} groups.

Phosphatic constituent--a portion of a *compound* made up of *phosphate.*

Photic zone--that part of the ocean in which there is sufficient penetration of light to support photosynthesis. The depth varies, but averages about 80 meters. Its lower boundary is the compensation depth. Also termed the euphotic zone.

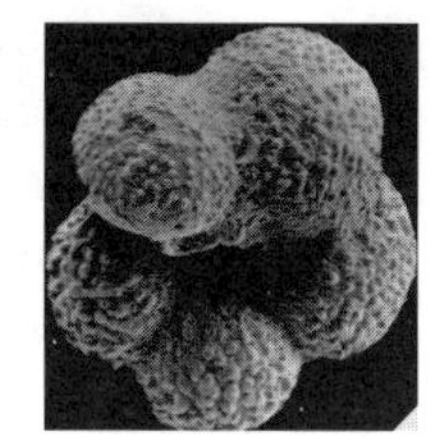

Planktic foram--a single celled organism with no tissue or organs which lives at or near the upper surface of the ocean characterized by the presence of a small shell (test) of one or many chambers composed of *calcite* or *silica*. Most *forams* are marine but freshwater forms are known.

Pleistocene epoch--an *epoch* of the *Quaternary period*, after the *Pliocene* of the *Tertiary* and before the *Holocene*, ranging from about 8 thousand to 2 million years ago, according to the *evolutionary time scale*.

Pliocene epoch--an *epoch* of the *Tertiary period*, after the *Miocene* and before the *Pleistocene*, ranging from about 2 to 5 million years ago, according to the *evolutionary time scale*.

Polar region--the geographic area in proximity to either or both of the two poles of the earth.

Pore water--subsurface water in the voids of a rock or *sediment*. Also termed interstitial water.

Postglacial--after or following a *glacial epoch*.

Precambrian era--all geologic time, and its corresponding rocks, before the beginning of the *Paleozoic*. Precambrian time has been divided according to several different systems, all of which use the presence or absence of evidence of life as a criterion.

Precession--the forward movement of a distinguishable feature in the *orbit* of one body about another; e.g., the westward migration of the *equinoxes* in earth's *orbit* around the sun with a complete rotation period of about 20,000 years.

Precipitation--the separation of a solute or condensate from a solution and falling in a gravitational field relative to the solution; e.g. rain, snow, or sleet falling from the atmosphere.

Productivity--a general term for the organic fertility of a body of water. The capacity of a lake or ocean to produce a particular organism.

Pteropod--*pelagic organisms* sometimes with shells (tests). The shells are generally conical and composed of *aragonite*. They are found from the *Cretaceous* to the present.

Quaternary Period--the second period of the *Cenozoic era*, following the *Tertiary*, ranging from the present to about 2 million years ago, according to the *evolutionary time scale*. It consists of two *epochs*: the *Pleistocene* and *Holocene*.

Radiolarian--a *protozoan* characterized by protoplasmic extensions radiating from the spheroidal main body composed mainly of a *siliceous* skeleton and living in a *pelagic* marine environment. Radiolaria are found from the *Cambrian* to the present.

Recrystallization--the formation of new crystalline mineral grains in a rock. The new grains are generally larger than the original grains, and may have the same or a different mineralogical composition. It is the way in which a deformed crystal aggregate releases stored strain energy due to deformation. Secondary recrystallization may follow, in which some grains grow very large by consuming neighboring grains.

Regional flood--the condition that occurs when water overflows the natural or artificial confines of streams or other bodies of water of regional geographic extent, or accumulates by drainage over low-lying areas. As opposed to *global flood* whereby the entire earth was covered by water.

Relict sediment--a sediment that had been deposited in equilibrium with its environment, but that is now unrelated to its present environment even though it remains unburied by later sediments; e.g., a land-laid or shallow-marine sediment occurring in deep water (as near the seaward edge of the *continental shelf*).

Reworked--said of a sediment, fossil, rock fragment, or other geologic material that has been removed or displaced by natural agents from its place of origin and incorporated in recognizable form in a younger formation, such as a "reworked tuff" carried by flowing water and redeposited in another locality.

Runoff--the water, derived from precipitation, that ultimately reaches stream channels and oceans.

Salinity--a measure of the quantity of dissolved salts in sea water. It is formally defined as the total amount of dissolved solids in sea water in parts per thousand by weight when all the carbonate has been converted to oxide, the bromide and iodide to chloride, and all organic matter is completely oxidized. In practice, salinity is not determined directly but is computed from chlorinity, electrical conductivity, refractive index, or some other property whose relationship to salinity is well established.

Sand--a rock fragment or detrital particle smaller than a granule and larger than a coarse silt grain, having a diameter in the range of 62 to 2000 *microns*. A loose aggregate of *unlithified mineral* or rock particles of sand size. The material is most commonly composed of quartz resulting from rock disintegration, and when the term "sand" is used without qualification, a *siliceous* composition is implied; but the particles may be of any *mineral* composition or mixture of rock or mineral fragments, such as "coral sand" consisting of *limestone* fragments.

Scientific results--the answer to a problem obtained by using the principles and methods of science; systematic and exact methods of observation, classification, hypothesis, theory, experimentation, and calculation, among others.

Sea-floor sediment--solid fragmental material that forms in layers on the ocean floor which is formed by weathering of rocks and is transported or deposited by air, water, or ice, or that accumulates by other natural agents, such as chemical precipitation from solution or secretion by organisms. It forms in layers at ordinary temperatures in a loose, unconsolidated form; e.g., sand, gravel, silt, mud, till, loess, or alluvium.

Seamount--an elevation of the sea floor, 1000 meters or higher, either flat-topped (called a guyot) or peaked (called a seapeak). Seamounts may be either discrete, arranged in a linear or random grouping, or connected at their bases and aligned along a ridge or rise.

Sediment layer--a distinct thickness of *sediment* which has similar consistency, properties, and appearance.

Sedimentary rock--a rock resulting from the consolidation of loose *sediment* that has accumulated in layers; or a chemical rock formed by precipitation from solution; or an organic rock consisting of the remains or secretions of plants and animals.

Sedimentology--the scientific study of sedimentary rocks and of the processes by which they were formed; the description, classification, origin, and interpretation of sediments.

Sediments--solid fragmental material that forms in layers on the floor of oceans or lakes which is formed by weathering of rocks and is transported or deposited by air, water, or ice, or that accumulates by other natural agents, such as chemical precipitation from solution or secretion by organisms.

Shelf sea--a shallow marginal sea situated on the continental shelf, rarely exceeding 275 meters in depth, e.g., the North Sea.

Sial--a petrologic name for the upper layer of the earth's crust, composed of rocks that are rich in silica and alumina.

Silica--a hard, glassy mineral found in a variety of forms, as in quartz, sand, opal, etc., made of SiO_2.

Siliceous constituent--a component in a rock or sediment containing silica.

Silurian period--a period of the *Paleozoic*, ranging from about 400 to 440 million years ago, according to the *evolutionary time scale*. The Silurian follows the *Ordovician* and precedes the *Devonian*. It is named after the Silures, a Celtic tribe.

Site--a geographical location in the oceans of the world where coring of sea floor sediments has been made by DSDP and ODP or other sea floor drilling projects.

Skeletal debris--pertaining to material derived from organisms and consisting of the hard parts secreted by the organisms or of the hard material around or within organic tissue. A term used to refer to a limestone that consists of skeletal matter (as distinguished from a fragmental limestone formed by mechanical transport).

Smearslide--a sample of sediment or other material "smeared" on a glass microscope slide by wiping the slide with the material.

Solar heating--the warming of the atmosphere or surface of the Earth by direct and indirect radiation from the sun. Indirect heating occurs by secondary scattering of the light from the atmosphere at angles away from the direction of the sun.

Solar radiation--the process by which electromagnetic radiation from the sun is propagated through space by virtue of joint undulatory variations in the electric and magnetic fields in space. This concept is to be distinguished from conduction and convection.

Solstice--either of the two points on the sun's apparent annual path where it is displaced farthest, north or south, from the earth's equator. Popularly, the time at which the Sun is farthest north or south; the "time of the solstice;" or when the sun "stands still."

Sonic velocity--the speed and direction of an acoustic wave, especially in water or sediments.

Sorting--the dynamic process by which sedimentary particles having some particular characteristic (such as similarity of size, shape, or specific gravity) are naturally selected and separated from associated but dissimilar particles by the agents of transportation (especially by the action of running water).

SPECMAP--a research project to conduct numerous spectral analyses on oxygen and carbon isotope records from sea floor sediment and correlate them with the harmonics of various orbital parameters of the earth/sun system.

Spectral analysis--the study of correlations between frequencies or periods of two or more signals.

Spud--to break ground with a drilling rig at the start of well-drilling operations.

Stable oxygen isotope data--measurements of the relative concentrations of oxygen isotopes that are not radioactive, such as ^{18}O and ^{16}O.

Stratification--the formation, accumulation, or deposition of material in layers.

Subantarctic--the geographical region near by but not in the Antarctic.

Submarine bank--a general term for limestone deposits that are locally abnormally thick and that appear to have formed over submerged shallow areas that rose above the general level of the surrounding sea floor. Submarine banks lack the hard, wave-resistant character of organic reefs.

Submarine ridge--an elongated, steep-sided elevation of the ocean floor, having rough topography.

Terrigenous--derived from the land or continent.

Tertiary Period--the first period of the *Cenozoic era* (after the *Cretaceous* of the *Mesozoic* era and before the *Quaternary*), ranging from about 2 to 65 million years ago, according to the *evolutionary time scale*.

Thermally-stratified ocean--the stratification of the ocean produced by changes in temperature at different depths and resulting in horizontal layers of differing densities.

Thermocline--a vertical, negative gradient of temperature that is characteristic of the layer of ocean water under the mixed layer; also, the layer in which this gradient occurs. A thermocline may be either seasonal or permanent.

Till--dominantly unsorted and unstratified drift, generally unconsolidated, deposited directly by and underneath a glacier without subsequent reworking by meltwater, and consisting of a heterogeneous mixture of clay, silt, sand, gravel, and boulders ranging widely in size and shape.

Topography--the general configuration of a land surface or any part of the earth's surface, including its relief and the position of its natural and man-made features.

Trial and error method--an accepted technique of finding a solution to a problem by guessing a solution, substituting into an equation, and then making a better estimate of the solution by observing a trend.

Triassic period--the first period of the *Mesozoic era* (after the *Permian* of the *Paleozoic era*, and before the *Jurassic*), ranging from 190 to 225 million years ago, according to the *evolutionary time scale*. The Triassic is so named because of its threefold division in the rocks of Germany.

Tropical--said of a climate characterized by high temperature and humidity and by abundant rainfall. An area of tropical climate borders that of equatorial climate.

Tropical sediment--sediments found on the sea floor which are thought to have lived in an ocean in a tropical climate.

Turbidite--a *sediment* or rock deposited from, or inferred to have been deposited from a *turbidity current*. It is characterized by *graded bedding* and moderate *sorting*.

Turbidity current--a density current in water, air, or other fluid, caused by different amounts of matter in suspension, such as a dry-snow avalanche or a descending cloud of volcanic dust; specifically, a bottom-flowing current laden with suspended sediment, moving swiftly (under the influence of gravity) down a subaqueous slope and spreading horizontally on the floor of the body of water.

Unconsolidated sediment--sediment which has not yet become firm or coherent rock. Consolidated sediment results in the gradual reduction in volume and increase in density of a soil mass from an increased load or effective compressive stress; e.g., the squeezing of fluids from pore spaces.

Uniformitarian process--the type of geological activity which operates to modify the Earth's crust and is assumed in the evolutionary model to have operated throughout geological time with essentially the same intensity and in the same regular manner as today. According to this view all past geological events can be explained by phenomena and forces observable today. The doctrine does not imply that all change is at a uniform rate, and does not exclude minor local catastrophes, but does exclude a catastrophic, global Flood.

Universal flood--the term recently applied to references in the Bible to a great flood which affected all the then-known world. This view teaches that the Genesis Flood was a *local flood* which covered the earth as far as the eye could see in the Tigris-Euphrates River valley, destroying the civilization and many animals and people, but not destroying the surface of the whole globe. This view is in opposition to a *Global Flood* which covered all the mountains of the entire globe and destroyed all air-breathing life except Noah and his family and animals on the Ark.

Upwelling--the rising of cold, heavy subsurface water toward the surface, especially along the western coasts of continents (as along the coast of southern California); the displaced surface water is transported away from the coast by the action of winds parallel to it or by diverging currents. Upwelling may also occur in the open ocean where cyclonic circulation is relatively permanent, or where southern trade winds cross the equator.

Uvigerina--a family of *forams* which are *marine*, *benthic*, and commonly found in cold waters; they live in the *abyssal*, as deep as 4500 meters.

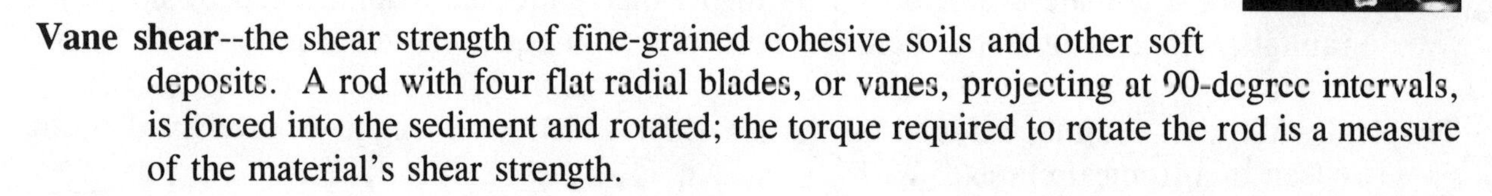

Vane shear--the shear strength of fine-grained cohesive soils and other soft deposits. A rod with four flat radial blades, or vanes, projecting at 90-degree intervals, is forced into the sediment and rotated; the torque required to rotate the rod is a measure of the material's shear strength.

Vertebrates--species characterized by an internal skeleton of cartilage or bone, and by specialized organization of the anterior end of the animal; the front of the body is a head that bears organs of sight, smell, taste, and hearing, and the front of the central nervous system is a brain.

Vibra-percussive corer--a coring device used to obtain a continuous sample from coarse-grained or semi-*lithified sediments*. The core barrel is supported by a tripod which rests on the *sediment* surface, and the barrel is driven into the *sediment* by a vibrator.

Weathering--the destructive process or group of processes by which earthy and rocky materials on exposure to atmospheric agents at or near the earth's surface are changed in color, texture, composition, firmness, or form, with little or no transport of the loosened or altered material; specifically, the physical disintegration and chemical decomposition of rock that produce an in situ mantle of waste and prepare sediments for transportation.

Working core--the half of a core which is available for destructive testing. Samples may be extracted for physical and chemical analysis.

X-ray diffraction--the breaking up of a beam of X-rays into dark or light bands on a photographic plate, usually by the three-dimensional periodic array of atoms in a crystal that has periodic repeat distances (lattice dimensions) of the same order of magnitude as the wavelength of the X-rays.

X-ray fluorescence--the emission of characteristic X-rays from a substance by exposure to X-rays of shorter wavelength.

X-ray mineralogy--the determination of the mineral composition of a substance by X-ray techniques.

Young-earth model--a conceptual or analytic representation of earth processes according to catastrophic events over thousands of years.

REFERENCES CITED

Aardsma, G.E., 1993. *Radiocarbon and the Genesis Flood.* Institute for Creation Research, San Diego, CA. 82 pp.

Arthur, M.A., W.E. Dean, and S.O. Schlanger. 1985. Variation in the global carbon cycle during the Cretaceous related to climate, volcanism, and changes in atmospheric CO_2. In *The Carbon Cycle and Atmospheric* CO_2*: Natural Variations Archean to Present*, ed. E.T. Sundquist and W.S. Broecker. pp 504-529. Washington D.C., American Geophysical Union, Geophysical Monograph 32.

Backman, J. 1979. Pliocene biostratigraphy of DSDP Sites 111 and 116 from the North Atlantic Ocean and the age of the Northern Hemisphere glaciation. *Stockholm Contributions to Geology*. 32: 115-137.

Barron, E.J. and J.M. Whitman. 1981. Oceanic sediments in space and time. In *The Sea, The Oceanic Lithosphere*, ed. C. Emiliani. 7: 689-731. John Wiley & Sons, New York.

Barron, E.J. 1983. A warm equable Cretaceous: the nature of the problem. *Earth Sci. Rev.* 19: 305-338.

Bowen, R. 1966. *Paleotemperature Analysis*. Elsevier, Amsterdam. 265 pp.

CLIMAP. 1976. The surface of the ice-age earth. *Science*. 191: 1131-1144.

CLIMAP. 1981. Seasonal reconstructions of the earth's surface at the last glacial maximum. *Geol. Soc. Amer.* Map and Chart Series No. 36.

Craig, H. and L. Gordon. 1965. Deuterium and oxygen 18 variations in the ocean and marine atmosphere. *Proc. Spoleto Conf. Stable Isotopes in Oceanographic Studies and Paleotemperatures*. pp. 9-130.

Craig, H. 1957. Isotopic standards for carbon and oxygen and correction factors for mass-spectrometric analysis of carbon dioxide. *Geochim. Cosmochim. Acta*. 12: 133-149.

Damuth, J.E. and R.W. Embley. 1981. Mass-transport processes on Amazon cone: Western equatorial Atlantic. *Bulletin of the American Association of Petroleum Geologists*. 65/4: 629.

Deines, P. 1970. Mass spectrometer correction factors for the determination of small isotopic composition variations of carbon oxygen. *Int. J. Mass. Spectrom. Ion Phys*. 4: 283.

Douglas, R.G. and S.M. Savin. 1971. Isotopic analyses of planktonic foraminifera from the Cenozoic of the Northwestern Pacific, Leg 6. In *Initial Reports of the Deep Sea Drilling Project*, A.G. Fisher et al. 6: 1123-1127. Washington D.C.: GPO.

Douglas, R.G. and S.M. Savin. 1973. *Oxygen and carbon isotope analyses of Cretaceous and Tertiary foraminifera from the Central North Pacific*. In *Initial Reports of the Deep Sea Drilling Project*, E.L. Winterer, J.I. Ewing et al, 17, 591-605, GPO, Washington, D.C.

Douglas, R.G. and S.M. Savin. 1975. Oxygen and carbon isotope analyses of Tertiary and Cretaceous microfossils from Shatsky Rise and other sites in the North Pacific Ocean. In *Initial Reports of the Deep Sea Drilling Project*, R.L. Larson, R. Moberly et al. 32: 509-520. Washington, D.C.:GPO.

Douglas, R.G. and S.M. Savin. 1977. Depth stratification in Tertiary and Cretaceous foraminifera based on oxygen isotope ratios. In *Marine Micropalaeo*.

Duplessy, J.C. 1972. *La geochimie des isotopes stables du carbone dans la mer*. Thesis, Univ. Paris VI. Paris. 196 pp.

Duplessy, J.C., C. LaLou, and A.C. Vinot. 1970. Differential isotopic fractionation in benthic foraminifera and paleotemperatures reassessed. *Science*. 168: 250-251.

Emiliani, C. 1954. Depth habitats of some species of pelagic foraminifera as indicated by oxygen isotope ratios. *Am. J. Sci*. 252: 149-158.

Emiliani, C. 1971. Depth habitats and growth stages of pelagic foraminifera. *Science*. 173: 1122-1124.

Epstein, S. and H.A. Lowenstam. 1953. Temperature-shell growth relations of recent and interglacial Pleistocene shoal-water biota from Bermuda. *J. Geol*. 51: 424-438.

Epstein, S. and T. Mayeda. 1953. Variation of ^{18}O content of waters from natural sources. *Geochim. Cosmochim. Acta* 4: 213-224.

Epstein, S., R. Buchsbaum, H. Lowenstam, and H.C. Urey. 1953. Revised carbonate-water isotopic temperature scale. *Geol. Soc. Am. Bull*. 64: 1315-1325.

Haq, B.U., J. Hardenbol, and P.R. Vail. 1987. Chronology of fluctuating sea levels since the Triassic (250 million years ago to present). *Science*. 235: 1156-1167.

Hays, J.D. and W.C. Pittman. 1973. Lithospheric plate motion, sea level changes, and climatic and ecological consequences. *Nature*. 246: 18-22.

Hays, J.D., J. Imbrie, and N.J. Shackleton. 1976. Variations in the Earth's orbit: Pacemaker of the Ice Ages. *Science*. 194: 1121-1132.

Hecht, A.D. and S.M. Savin. 1972. Phenotypic variation in oxygen isotope ratios in recent planktonic foraminifera. *J. Foraminiferal Res*. 2: 55-67.

Hodell, D., D. Williams, and J.P. Kennett. 1985. Late Pliocene reorganization of deep vertical water mass structure in the western South Atlantic: Faunal and isotopic evidence. *Bull. Geol. Soc. Amer*. 96: 495-503.

Jacobs, G.A., H.E. Hurlburt, J.C. Kindle, E.J. Metzger, J.L. Mitchell, W.J. Teague, and A.J. Wallcraft. 1994. Decade-scale trans-Pacific propagation and warming effects of an El Niño anomaly. *Nature*. 370: 360-363.

Keigwin, L.D. 1979. Late Cenozoic stable isotope stratigraphy and paleoceanography of DSDP sites from the east equatorial and central North Pacific Ocean. *Earth Planet. Sci. Letters*. 45: 361-382.

Keigwin, L.D. 1986. North Atlantic record of late Neogene climatic change. *S. Afr. J. Sci*. 82: 499-500.

Kennett, J.P., R.E. Houtz, P.B. Andrews, A.R. Edwards, V.A. Gostin, M. Hajós, M. Hampton, D.G. Jenkins, S.V. Margolis, A.T. Ovenshine, K. Perch-Nielsen. 1977. Descriptions of Procedures and Data for Sites 277, 279, and 281 by the Shipboard Party. In *Initial Reports of the Deep Sea Drilling Project*. 29: 45-58, 191-202, and 271-285. GPO: Washington, D.C.

Kennett, J.P. 1977. Cenozoic evolution of Antarctic glaciation, the circum-Antarctic ocean, and their impact on global palaeoceanography. *J. Geophys. Res*. 82: 3843-3860.

Lidz, B., A. Kehm, and H. Miller. 1968. Depth habitats of pelagic foraminifera during the Pleistocene. *Nature*. 217: 245-247.

Martinson, D., N.G. Pisias, J.D. Hays, J. Imbrie, T.C. Moore, and N.J. Shackleton. 1987. Age dating and the orbital theory of the ice ages: development of a high resolution 0 to 300,000-year chronology. *Quat. Res*. 27: 1-29.

Matthews, R.K. and R.Z. Poore. 1980. Tertiary delta ^{18}O record and glacio-eustatic sealevel fluctuations. *Geology*. 8: 501-504.

Milankovitch, M. 1930. Mathematische klimalehre und astronomische theorie der Klimaschwankungen. In *Handbuch der Klimatologie*, I.W. Koppen and R. Geiger (Eds.). Gebruder Borntraeger, Berlin.

Milankovitch, M. 1941. Canon of insolation and the ice age problem (in Yugoslavian). *Serb. Acad. Beorg.*, Spec. Publ. 132. English translation by Israel Program for Scientific Translations, Jerusalem, 1969.

Miller, K.G., R.G. Fairbanks, and G.S. Mountain. 1987. Tertiary oxygen isotope synthesis, sea level history and continental margin erosion. *Paleoceanography*. 1: 1-20.

Morris, H.M. 1976. *The Genesis Record.* Baker Book House, Grand Rapids, Michigan and Creation-Life Publishers, San Diego, California.

Morris, H.M. 1996. The Geologic Column and the Flood of Genesis. *Creation Research Society Quarterly* (In Press).

Oard, M.J. 1990. *An Ice Age Caused by the Genesis Flood.* Institute for Creation Research Monograph, San Diego, California, 243 pp.

Oba, T. 1969. Biostratigraphy and isotopic paleotemperature of some deep-sea cores from the Indian Ocean. *Sci. Rep. Tohoku Univ. Sendai 2nd Ser. (Geol.)*. Vol. 41, No. 2: 129-195.

Oberhansli, H. and K.J. Hsu. 1986. Paleocene-Eocene Paleoceanography. In *Mesozoic and Cenozoic Oceans*, ed. K.J. Hsu. pp 85-100. Washington D.C., American Geophysical Union Geodynamics Series 15.

Oberhansli, H. 1986. Latest Cretaceous-early Neogene oxygen and carbon isotopic record at DSDP sites in the Indian Ocean. *Marine Micropalaeo*. Special issue on stable isotopes, 10: 91-116.

Palmer, A.R. 1983. The Decade of North American Geology 1983 Geological Time Scale. *Geology*. 11: 503-504.

Poore, R.Z. and R.K. Matthews. 1984. Late Eocene-Oligocene oxygen and carbon isotope record from the South Atlantic DSDP site 522. *Init. Repts.*, DSDP73, pp 725-735.

Prell, W.L. 1984. Covariance patterns of foraminiferal $\delta^{18}O$: an evaluation of Pliocene ice volume near 3.2 Ma., *Science*. 226: 692-693.

Revelle, R.R. 1944. Marine bottom sediment collected in the Pacific Ocean by the Carnegie on its seventh cruise. In *Carnegie Institution Washington Publ. 556*, part 1, 180 pp.

Roth, A.A. 1985. Are millions of years required to produce biogenic sediments in the deep ocean? *Origins*. 5: 48-56.

Savin, S.M. and F.G. Stehli. 1974. Interpretations of oxygen isotope paleotemperature measurements: effect of the $^{18}O/^{16}O$ ratio of sea water, depth stratification of foraminifera, and selective solution. Colloq. Int. CNRS No. 219. In *Les Methodes Quantitative d'Etude des Variations au cours du Pleistocene*. pp 183-191.

Savin, S.M., R.G. Douglas. 1973. Stable isotope and magnesium geochemistry of recent planktonic foraminifera from the South Pacific. *Geol. Soc. Am. Bull.* 84: 2327-2342.

Savin, S.M., R.G. Douglas, F.G. Stehli. 1975. Tertiary marine paleotemperatures. *Geol. Soc. Am. Bull.* 86: 1499-1510.

Shackleton, N.J. et al. 1984. Oxygen isotope calibration of the onset of ice-rafting and history of glaciation in the North Atlantic region. *Nature*. 307: 620-623.

Shackleton, N.J. and J.P. Kennett. 1975. Paleotemperature history of the Cenozoic and the initiation of Antarctic glaciation: oxygen and carbon isotope analyses in DSDP sites 277, 279, and 281. In *Initial Reports of the Deep Sea Drilling Project*. 29: 743-755. GPO: Washington, D.C.

Shackleton, N.J. 1968. Depth of pelagic foraminifera and isotopic changes in Pleistocene oceans. *Nature*. 218: 79-80.

Shackleton, N.J. 1974. Attainment of isotopic equilibrium between ocean water and benthonic foraminifera genus Uvigerina: isotopic changes in the ocean during the last glacial. Colloq. Int. CNRS No. 219, *Les Methodes Quantitative d'Etudes des Variations du Climat au cours du Pleistocene*. pp 203-209.

Shackleton, N.J., J.D.H. Wiseman, and H.A. Buckley. 1973. Nonequilibrium isotopic fractionation between seawater and planktonic foraminiferal tests. *Nature*. 242: 177-179.

Spaeth, C., J. Hoefs, and U. Vetter. 1971. Some aspects of isotopic composition of belemnites and related paleotemperatures. *Geol. Soc. Am. Bull.* 82: 3139-3150.

Stevens, G.R., R.N. Clayton. 1971. Oxygen isotope studies on Jurassic and Cretaceous belemnites from New Zealand and their biogeographic significance. *J. Geol. Geophys.* 14: 829-897.

Urey, H.C., H.A. Lowenstam, S. Epstein, and L.R. McKinney. 1951. Measurement of paleotemperatures and temperatures of the upper Cretaceous of England, Denmark, and the southeastern United States. *Geol. Soc. Am. Bull.* 62: 399-416.

Ussher, J. 1786. *The Annals of the World*. Printed by E. Tyler for J. Crook and G. Bedell, London.

Van Donk, J. 1970. *The oxygen isotope record in deep sea sediments.* PhD thesis, Columbia Univ., New York. 228 pp.

Vardiman, L. 1993. *Ice Cores and the Age of the Earth.* ICR Monograph, Institute for Creation Research, San Diego, CA, 73 pp.

Vergnaud Grazzini, C. 1980. Cenozoic paleotemperatures at Site 398, eastern North Atlantic; diagenetic effects on carbon and oxygen isotope signal. *Init. Repts. DSDP47*, pp 507-511. Washington D.C.: GPO.

Weissert, H., J.A. McKenzie, R.C. Wright, M. Clark, H. Oberhansli, and M. Casey. 1984. Paleoclimatic record of the Pliocene at DSDP Sites 519, 521, 522, and 523 (Central South Atlantic). *Init. Repts. DSDP73*, pp 701-715. Washington D.C.: GPO.

Williams, D.F., N. Healy-Williams, R.C. Thunell, and A. Leventer. 1983. Detailed stable isotope and carbonate records from the upper Maastrichtian-lower Paleocene of Hole 516F (Leg 72) including the Cretaceous/Tertiary boundary. *Init. Repts. DSDP72*, pp 921-930. Washington D.C.: GPO.